Document-Based Cases for Technical Communication

Second Edition

Roger Munger
Boise State University

Bedford/St. Martin's   *Boston • New York*

For Bedford/St. Martin's

Senior Executive Editor: Leasa Burton
Developmental Editor: Regina Tavani
Senior Production Editor: Gregory Erb
Production Supervisor: Samuel Jones
Senior Marketing Manager: Molly Parke
Copy Editor: Judith Riotto, Ganymede Editorial Services
Permissions Manager: Kalina K. Ingham
Senior Art Director: Anna Palchik
Text Design and Composition: Books By Design, Inc.
Cover Design: Marine Miller
Printing and Binding: Malloy Lithographing, Inc.

President, Bedford/St. Martin's: Denise B. Wydra
Presidents, Macmillan Higher Education: Joan E. Feinberg and Tom Scotty
Editor in Chief: Karen S. Henry
Director of Marketing: Karen R. Soeltz
Production Director: Susan W. Brown
Associate Production Director: Elise S. Kaiser
Managing Editor: Elizabeth M. Schaaf

Manufactured in the United States of America.

For information, write: Bedford/St. Martin's, 75 Arlington Street, Boston, MA 02116   (617-399-4000)

ISBN 978-1-4576-1502-3

Acknowledgments
Photo in Document 2.5: *Transverse Ladybeetle*. Whitney Cranshaw, Colorado State University, Bugwood.org.
The illustrations in Document 4.6 are from http://dps.sd.gov/enforcement/highway_safety/pedestrian _and_bikesafety.aspx and www.monroecounty.gov/safety-faq.php.
The information in Document 5.12 was obtained from www.eia.gov/renewable/annual/trends/pdf /trends.pdf.

Preface for Instructors

Sample documents are at the heart of the technical communication course, embodying the many ways that writers respond to complex writing situations. *Document-Based Cases for Technical Communication* aims to provide students with an even richer understanding of rhetorical situations by presenting clusters of related documents, along with realistic tasks, in the context of workplace scenarios. This book provides seven cases that ask students to analyze and produce common workplace documents, such as business graphics, definitions, memos, e-mails, proposals, technical reports, instructions, and presentation graphics.

The documents in this book are not meant to serve as models of the "perfect" memo, proposal, or e-mail. Instead, these documents represent the kinds of raw materials that students are likely to encounter and be asked to work with in real writing situations: examples and templates from a particular workplace, documents that need to be revised, directions from supervisors, and informal questions and advice from colleagues. With significant background information and guidance about how to address the challenges, each case offers students multiple opportunities to make decisions about audience and purpose, to see how their decisions affect the documents they develop, and ultimately to imagine how they might apply course principles and concepts to their chosen career.

New to This Edition

We've reimagined the second edition for the digital age, integrating online communication, social media, and multimodal writing at every opportunity. Four of the seven cases are new, giving students the chance to immerse themselves in rhetorical situations that best mirror the kinds of workplaces they will enter in the twenty-first century. And in both new and updated cases, 20 new tasks give students the opportunity to create and improve documents in a variety of new genres, including blogs and microblogs, online training modules, and press releases. We've placed greater emphasis on document design in the second edition as well, offering fresh opportunities for students to develop their own documents, many of which must be designed specifically for the Web. A new, more Web-oriented text design reflects and reinforces these updates in content. For instructors, new scoring guides are available on the book's companion Web site to help you evaluate students' performance on each case.

Key Features

Each of the seven cases includes the following features:

- Four realistic workplace tasks require students to analyze sample documents, use information to solve a problem, and create common documents.

- Background information provides a rhetorically rich context in which to work with the sample documents and materials.

- Advice for meeting the presented challenges helps students focus on the best ways to begin and complete the tasks.

- Concluding activities ask students to reflect on the case scenarios and to apply what they have learned to other workplace situations.

- Full color electronic copies of case documents are available for students to download and work with at **bedfordstmartins.com/techdocs**.

- Grading rubrics for instructors offer criteria for evaluating student responses to each of the case tasks; these scoring guides are available in an instructor-only area of the student site, located at **bedfordstmartins.com/techdocs/instructor**.

Advice for Using This Book

Although this book could serve as a stand-alone text, it is best used in combination with a full-length technical or business communication text. To package *Document-Based Cases for Technical Communication* for free with Mike Markel's *Technical Communication*, Tenth Edition, use ISBN 978-1-4576-1577-1. *Document-Based Cases for Technical Communication* is also available as an e-book within *TechCommClass*, a unique online course space for technical communication. Visit yourtechcommclass.com to learn more.

Because each case provides multiple sample documents and several opportunities for students to interact with the materials, you can adapt the cases to meet your course's objectives and your students' needs. You might wish to consider the following strategies for using this casebook in your course:

- Build a one- to two-week unit around a particular case, or build an entire course around the seven workplace genres featured in the cases.

- Use case documents as discussion starters.

- Ask students to respond to one of the shorter tasks as an in-class or take-home quiz.

- Assign selected tasks as homework to reinforce in-class learning.

- Use one or more of a case's tasks as a major assignment in your course.

- Create variations of the cases by changing the audience, purpose, or context for a task.

Overall, this casebook provides you with flexible ways to easily supplement your regular text with in-depth discussions of realistic workplace situations and documents.

Acknowledgments

I'd like to thank those instructors who took the time to review the first edition of this text and suggest a multitude of changes and improvements for the second: Michael Dittman, Butler County Community College; Kristin Johnson, Metropolitan State University; Mark Ristroph, Augusta Technical College; Sherry Robertson, Arizona State University; Patricia Scharf, Metropolitan State University; Bruce Wehler, Pennsylvania College of Technology; and Stephanie Zerkel-Humbert, Maple Woods Community College.

Many thanks also to the reviewers who thoughtfully responded to early drafts and contributed good ideas to the first edition of this book: Bruce Brandt, South Dakota State University; William FitzGerald, University of Maryland, College Park; William Garcia, University of Maryland, College Park; David Gaskill, Saginaw Valley State University; Anne Lehman, Milwaukee Area Technical College; and Lisa DuPree McNair, Georgia Institute of Technology.

Thanks to Joan Feinberg, Denise Wydra, and Karen Henry for their encouragement and support. Thanks to Leasa Burton for asking me to take on this project and for working with me to develop the ideas and content for the first edition. Thanks to Regina Tavani for her editorial support and advice. I am grateful to Janis Owen for a remarkable text design that brings the content of the project to life in black and white. Thanks also to Anna Palchik for her design direction and to Gregory Erb for his meticulous attention during production.

I am also grateful to my wife, Lisa, who supported and encouraged me throughout the project. Lisa provided specific details, documents, and technical advice for the Definitions and Descriptions and Instructions cases. She also made valuable suggestions for improving the other cases.

Roger Munger

Introduction for Students

New graduates making the transition from college to the workplace are often surprised by the amount of writing they are asked to do—especially those graduates who did not major in disciplines typically associated with writing. New hires also report that their supervisors expect them to already have the writing skills necessary to communicate complex information in a clear and understandable manner to a variety of readers. Not surprisingly, new hires also need to quickly learn the preferred form and style of an organization's documents. This book provides you with a head start on refining the writing skills that will help you adapt to different expectations in the workplace.

Anatomy of a Case

A case is a detailed story of a workplace communication problem, complete with characters, dialogue, and props in the form of sample documents. Based on an actual problem solved by practicing technical communicators, each case in this book provides a realistic workplace situation, a series of documents related to the situation, and several tasks for you to complete. Together, these elements offer snapshots of what it is like to create documents that solve problems for readers in the workplace.

Effective technical communication addresses particular readers and helps them solve problems. That is, the decisions you make when creating a technical document depend on the intended audience and purpose. Cases give you opportunities to explore how the complexities of specific workplace situations shape technical documents.

The elements of each case in this book work together to present a complex writing situation.

- **The situation** section provides details about the workplace scenario and writing situation. Use these details to help make decisions about the audience and purpose for your writing. For some cases, the situation includes model documents and additional examples.

- **The challenge** describes the difficult issues that you will need to consider as you solve the case's communication problems.

- **Your job** describes your general role in the workplace scenario and lists the activities you will need to perform to complete the case.

- **Four tasks** present your writing assignments, each with a set of related documents. Each task begins with an introduction that explains the communication problem, describes your audience and purpose, and specifies the type of document you must create.

- The **documents** you will need for each task appear in the shaded section that begins below the task introduction and may continue across several pages. You can download all of these documents from **bedfordstmartins.com/techdocs**.

- **When you're finished** with one or more tasks in a case, your instructor may ask you to reflect on the skills you have learned or use those skills to solve similar problems.

Advice on Getting Started

Just like the communication challenges you will likely face in your chosen career, the solutions to the problems raised in these cases might not be readily apparent, and there is often not a single "correct" answer. When responding to tasks, consider the following advice:

- If you do not understand a particular word or concept, carry out some research before continuing with the task. An important skill for you to develop is the ability to locate relevant resources, learn about a subject quickly, and explain it to someone else.

- Read the sample documents carefully and more than once. Unfamiliar content will make seemingly simple documents a challenge.

- Consider several possible approaches to the problem before deciding on a course of action. Often, your first idea is not the most effective solution.

- When faced with a decision, use your understanding of the document's audience and your purpose as a guide. These two basic elements of the writing situation will determine your document's design, content, and tone.

- For each decision you make, be able to explain what you did and why you did it. Being able to articulate your choices allows you to demonstrate that your documents are based on sound technical communication principles.

- Be creative in your problem solving, and have fun with these cases.

Finally, think of these cases as a way to practice in a nonthreatening environment the single greatest determining factor of success in the workplace: your communication skills. Launching your career with a firm command of these skills in place will ensure you are well prepared to face the unique challenges of today's workplace.

Page Design for Assignments

Each case presents **four tasks**, or assignments, with a set of related documents. The task begins with an introduction that explains the communication problem, describes your audience and purpose, and specifies what type of document you must create.

The **documents** you will need to complete each task appear in the shaded section that begins below the task introduction and may continue across several pages. Digital files for the documents can be found at **bedfordstmartins.com/techdocs**.

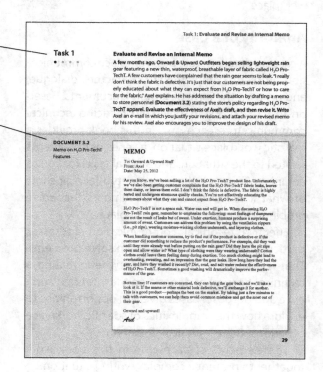

Contents

Preface for Instructors iii
Introduction for Students v

Case 1	**Graphics:** Selecting and Presenting Data 1

The Situation 1

FIGURE 1.1 Sample APR Calculations for a Women's Golf Team 2

The Challenge 2

Your Job 3

When You're Finished 3

Task 1   **Evaluate the Effectiveness of Graphics** 4

DOCUMENT 1.2 Line Graph of Academic Achievement Trends 4
DOCUMENT 1.3 Bar Graph of Academic Achievement for All Women's Sports 5

Task 2   **Assess and Improve Communication Strategies** 6

DOCUMENT 1.4 APR Statistics for Women's Sports, 2009–2010 6
DOCUMENT 1.5 APR Statistics for Men's Sports, 2009–2010 7
DOCUMENT 1.6 APR Snapshot 8

Task 3   **Create Effective Graphics and Speaker's Notes** 9

DOCUMENT 1.7 Average Multi-year APR Scores, 2009–2010 10

Task 4   **Write a Press Release, and Integrate Text and Graphics** 11

DOCUMENT 1.8 Press Release Outline 12

Case 2	**Definitions and Descriptions:** Helping Your Readers Understand 15

The Situation   15

    FIGURE 2.1 Bayside's Community Garden   15

    FIGURE 2.2 Model Parenthetical and Sentence Definitions   16

The Challenge   16

Your Job   17

When You're Finished   17

Task 1   **Analyze Parenthetical and Sentence Definitions**   18

    DOCUMENT 2.3 Parenthetical Definitions   18

    DOCUMENT 2.4 Sentence Definitions   19

Task 2   **Evaluate and Revise Descriptions**   20

    DOCUMENT 2.5 Object Description of Beneficial Insects   20

    DOCUMENT 2.6 Mechanism Description of Garden Tiller   21

    DOCUMENT 2.7 Process Description of Erosion Control   22

    DOCUMENT 2.8 Object Description of USDA Plant Hardiness Zone 8b   23

Task 3   **Clarify Definitions with Graphics**   24

Task 4   **Write Definitions**   25

    DOCUMENT 2.9 E-mail with Gardening Terms to Be Defined   25

Case 3	**Correspondence:** Considering Your Reader's Point of View 27

The Situation   27

    FIGURE 3.1 Axel Geirsson   27

The Challenge   27

Your Job   28

When You're Finished   28

Task 1   **Evaluate and Revise an Internal Memo**   29

    DOCUMENT 3.2 Memo on H_2O Pro-TechT Features   29

Task 2   **Evaluate and Revise a Response to a Customer**   30

    DOCUMENT 3.3 Customer Claim Letter   30

    DOCUMENT 3.4 Response to Customer Claim Letter   31

Task 3   **Respond to Customer Claims**   32

    DOCUMENT 3.5   Customer Claim E-mail 1   32
    DOCUMENT 3.6   Customer Claim E-mail 2   33

Task 4   **Write Microblogs and a Blog Post**   34

    DOCUMENT 3.7   Microblog Posts Critical of Company   34
    DOCUMENT 3.8   Blog Post Critical of Company   35

Case 4   **Proposals:** Seeing Proposals through Reviewers' Eyes   37

The Situation   37
    FIGURE 4.1   Sample Funded Proposal   38
The Challenge   39
Your Job   39
When You're Finished   39

Task 1   **Evaluate Proposals**   40

    DOCUMENT 4.2   Skyview Coalition Proposal   40
    DOCUMENT 4.3   Gameday Brigade Proposal   41
    DOCUMENT 4.4   Teen Initiative Plan Proposal   42
    DOCUMENT 4.5   Defensive Driving Course Proposal   43

Task 2   **Reflect on the Review Process**   44

Task 3   **Respond to Proposals**   44

Task 4   **Write a Proposal**   45

    DOCUMENT 4.6   E-mail with Notes on Bicycle Helmet Program   45

Case 5   **Reports:** Learning to Write in an Organization   47

The Situation   47
The Challenge   47
    FIGURE 5.1   Cover of the EIA's *Web Editorial Style Guide*   48
Your Job   48
When You're Finished   49

Task 1 **Learn about an Organization's Reports 50**

FIGURE 5.2 "About EIA" Page of the EIA Web Site 50
FIGURE 5.3 "Mission and Overview" Page of the EIA Web Site 51
FIGURE 5.4 "EIA Offices" Page of the EIA Web Site 52
DOCUMENT 5.5 Preface from *International Energy Outlook 2010* 53

Task 2 **Compare and Evaluate Report Designs 55**

DOCUMENT 5.6 Web Page of Most Recent *International Energy Outlook* Report 55
DOCUMENT 5.7 Web Page of Archived *International Energy Outlook 2010* 56
DOCUMENT 5.8 Highlights Page from Printed Version of Archived *International Energy Outlook 2010* 57

Task 3 **Learn about Appropriate Report Content 58**

DOCUMENT 5.9 Page from *Greenhouse Emissions* Report 58
DOCUMENT 5.10 Excerpt on Energy-Related Carbon Dioxide Emissions from *Greenhouse Emissions* Report 59
DOCUMENT 5.11 Excerpt on Residential-Sector Carbon Dioxide Emissions from *Greenhouse Emissions* Report 60

Task 4 **Use an Organization's Style Guide 61**

DOCUMENT 5.12 E-mail Query on How to Develop a Preface 61
DOCUMENT 5.13 E-mail Query on How to Format a Table 62

Case 6 **Instructions: Guiding Readers in Performing a Task 63**

The Situation 63
FIGURE 6.1 Engineer's Sketch of ECG Lead II Waveform Display 63
The Challenge 63
Your Job 64
When You're Finished 64

Task 1 **Plan an Interview with a Subject-Matter Expert 65**

DOCUMENT 6.2 E-mail Requesting Information on the Priti4.3 Monitor 65

Task 2 **Evaluate Instructions 66**

FIGURE 6.3 Engineer's Sketch of Monitor User Interface 66
DOCUMENT 6.4 Basic Instructions for Operating Priti5 Monitor 67

Task 3   **Write Instructions**   **68**

DOCUMENT 6.5   Notes for Operating Priti5 Monitor   **68**

Task 4   **Develop an Online Training Module**   **71**

Case 7   **Presentation Graphics:** Highlighting Important Information   **73**

The Situation   **73**

FIGURE 7.1   Sample E-mail Announcing Open Enrollment Meeting   **74**

The Challenge   **74**

Your Job   **74**

When You're Finished   **75**

Task 1   **Design Presentation Slides**   **76**

DOCUMENT 7.2   Opening Slide for Presentation   **76**

DOCUMENT 7.3   Presentation Slide with Agenda   **77**

DOCUMENT 7.4   Presentation Slide on the Paper Application   **78**

Task 2   **Present Information Visually**   **79**

DOCUMENT 7.5   Presentation Slide with Pie Chart   **79**

DOCUMENT 7.6   Presentation Slide on Survey Results   **80**

DOCUMENT 7.7   Presentation Slide on Benefits   **81**

Task 3   **Create an Online Presentation**   **82**

DOCUMENT 7.8   Presentation Notes   **83**

Task 4   **Prepare a Handout**   **84**

DOCUMENT 7.9   E-mail with Information for Presentation Handout   **84**

Graphics
Selecting and Presenting Data

The Situation

The Gulls Department of Athletics at Bayside State College fields 10 National Collegiate Athletic Association (NCAA) Division I intercollegiate teams: five women's sports (volleyball, tennis, golf, soccer, and basketball) and five men's sports (tennis, golf, soccer, basketball, and water polo). The teams compete in the Metro Sports Conference. In addition to tracking the graduation rates and grade point averages of their student-athletes, the staff at the Department of Athletics uses academic progress rates (APRs) developed by the NCAA to measure the academic achievement of student-athletes at Bayside.

A team's APR score reflects how successful its student-athletes are at maintaining their academic eligibility and whether they remain enrolled in school. To be academically eligible to compete, student-athletes must meet a certain grade point average and make steady progress toward their degree by passing a specific number of degree-applicable courses each semester. Because turnover of nonscholarship players on a team roster can be high and because these student-athletes have not signed an agreement to stay on the team or at the school, coaches have less influence over them. Consequently, the NCAA includes only students receiving athletically related financial aid when calculating APR scores.

Recently, Bayside calculated two APR scores for each team: a *multi-year* APR score that reflects academic performance and progress over four years and a *single-year* score that focuses on just one academic year. Although a multi-year score includes data from the past four academic years, it is referred to by the most recent academic year included. Similarly, a single-year score is labeled by the academic year from which the data was drawn. For example, a multi-year score for 2009–2010 includes data from fall 2009, spring 2010, and the three previous years, going back to fall 2006. A single-year score for 2009–2010 includes data only from fall 2009 and spring 2010.

The APR measures the academic achievement of teams each fall and spring using a point system (**Figure 1.1**). Each student-athlete on scholarship earns one point if he or she stays in school and one point for being eligible to compete each term. Thus, a scholarship student-athlete can earn up to four points each year: two in the fall and two in the spring. A team's *single-* and *multi-year* scores are calculated by dividing the total points earned (over one year or the past four years) by the total points possible and then multiplying the result by 1,000. For example, a team earns 34 points for the year out of a possible 36. The single-year

1

Figure 1.1
Sample APR Calculations
for a Women's Golf Team

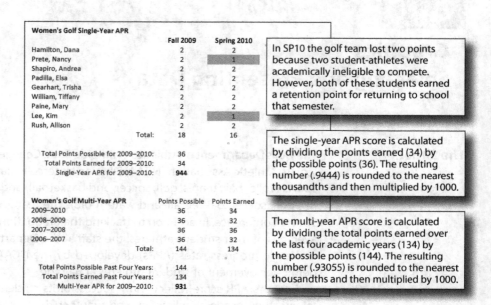

Women's Golf Single-Year APR

	Fall 2009	Spring 2010
Hamilton, Dana	2	2
Prete, Nancy	2	1
Shapiro, Andrea	2	2
Padilla, Elsa	2	2
Gearhart, Trisha	2	2
William, Tiffany	2	2
Paine, Mary	2	2
Lee, Kim	2	1
Rush, Allison	2	2
Total:	18	16

Total Points Possible for 2009–2010:	36
Total Points Earned for 2009–2010:	34
Single-Year APR for 2009–2010:	**944**

Women's Golf Multi-Year APR

	Points Possible	Points Earned
2009–2010	36	34
2008–2009	36	32
2007–2008	36	36
2006–2007	36	32
Total:	144	134

Total Points Possible Past Four Years:	144
Total Points Earned Past Four Years:	134
Multi-Year APR for 2009–2010:	**931**

In SP10 the golf team lost two points because two student-athletes were academically ineligible to compete. However, both of these students earned a retention point for returning to school that semester.

The single-year APR score is calculated by dividing the points earned (34) by the possible points (36). The resulting number (.9444) is rounded to the nearest thousandths and then multiplied by 1000.

The multi-year APR score is calculated by dividing the total points earned over the last four academic years (134) by the possible points (144). The resulting number (.93055) is rounded to the nearest thousandths and then multiplied by 1000.

APR score for the team is 944 (34 divided by 36 = .944, which is then multiplied by 1,000). A perfect score for the year is 1,000. Teams with low APR scores face penalties, such as scholarship losses and restrictions on practice and competition.

Because of your current position as an intern with the Bayside State College Sports Information Office and your expertise as a technical communicator, the athletic director, Cheryl Mueller, has enlisted your help in preparing materials to communicate the recently released APR scores. In particular, she seeks your assistance in explaining APR scores, selecting relevant APR data, and presenting the information in ways that coaches, faculty, and community members will understand. Because APR scores track each team's classroom performance, the scores can be used to gauge the commitment of the student-athletes and their coaches to academics, to measure a team's academic improvement, and to compare a team's academic performance to that of conference rivals. An athletic department with consistently high APR scores demonstrates that its teams are competitive in the classroom as well as on the field.

The Challenge

Organizations often collect a wide variety of data on organizational performance, and they manipulate these data in numerous ways. This results in a mass of information that must be organized and transformed before the data can be used for a specific purpose. The data-collection process may seem foreign at first, and you may need to do further research at the organization or on the Internet to understand the numbers. Your challenge is to avoid becoming overwhelmed by the amount of data you must analyze and to instead discover ways to identify, manage, and present the data that are applicable to your task while ignoring what is irrelevant.

Your Job

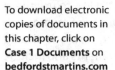

To download electronic copies of documents in this chapter, click on **Case 1 Documents** on **bedfordstmartins.com /techdocs**.

The data that Department of Athletics staff members have collected can be confusing. Your job is to help organize and present the data to show the faculty and larger community that the department is focused on the academic success of its student-athletes as well as to help coaches and athletic administrators track academic performance and, if necessary, identify teams struggling in the classroom. This case asks you to analyze data on the academic achievement of several intercollegiate teams at a state college. You may be asked to do the following:

▶ Evaluate the effectiveness of graphics to be used in a report published by the Department of Athletics.

▶ Revise documents the department currently uses to communicate academic achievement scores to coaches.

▶ Create graphics that highlight various teams' academic achievement to the faculty senate.

▶ Create graphics and text for a press release that communicates the Department of Athletics' recent academic achievement to various stakeholders, such as campus administrators, faculty, and fans.

Your instructor will tell you which of the tasks you are to complete.

Get started on your job.

When You're Finished

Reflecting on This Case In a 250- to 500-word response to your instructor, discuss (a) what you learned from this case, (b) how you could relate this case to work situations you will face in your chosen career, and, if applicable, (c) the ways in which this case compares to similar situations you have already faced at work. Your instructor will tell you whether your response should be submitted as a memo, an e-mail, or a journal entry, or in a different format.

Moving beyond This Case Collect data for two different products within a specific category. For example, you might choose to research the nutritional values of two different energy bars, or you might investigate the features of two different video game consoles, the cost of living in two different towns, or the value of two different stocks. Using the data you collect, create the following:

▶ A graphic effectively showing that option 1 is the best choice.

▶ A graphic effectively showing that option 2 is the best choice.

▶ An unbiased graphic comparing the two options.

Be prepared to explain the choices you made when selecting and presenting data as well as the rhetorical situations in which each of your graphics would be appropriate.

Task 1

● ● ● ●

Evaluate the Effectiveness of Graphics

Cheryl Mueller, the athletic director, shows you two graphics (**Documents 1.2** and **1.3**) she plans to use in the department's next annual report. "I haven't had a chance to ask anyone about these graphics," she tells you. "I'd like to use these graphics in our next report. What do you think?" Write Cheryl a brief memo evaluating the effectiveness of the two graphics and suggesting how she should revise them. You may wish to annotate the graphics with comments and include them with your memo.

DOCUMENT 1.2

Line Graph of Academic Achievement Trends

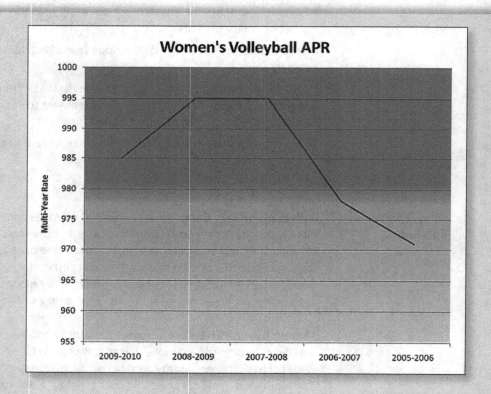

DOCUMENT 1.3

Bar Graph of Academic
Achievement for All
Women's Sports

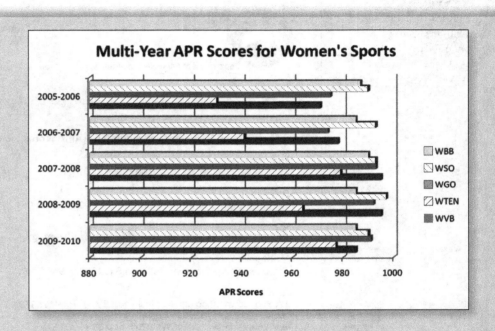

Task 2

● ● ● ●

Assess and Improve Communication Strategies

At the fall coaches' meeting, attendees receive the latest APR statistics (**Documents 1.4** and **1.5**) as well as a sheet briefly explaining APR scores (**Document 1.6**). An athletics staff member sitting next to you points to the sheets and asks, "You know something about graphics and design, don't you? Do you think we could improve the way we communicate this information? To be honest, I'm

DOCUMENT 1.4

APR Statistics for Women's Sports, 2009–2010

Bayside State

Department of Intercollegiate Athletics   7463 University Drive, Bayside, California 95524-1025

phone (707) 555-9175
fax (707) 555-1633
www.gullsports.com

Most Recent Multi-year and single-Year APR Scores for Women's Sports

Basketball
MY: 985, no change; SY: 982; MMY: 963

Golf
MY: 991,* down 1 point over last year; SY: 1000;* MMY: 981

Soccer
MY: 990,* down 7 points over last year; SY: 971; MMY: 978

Tennis
MY: 977, up 13 points over last year; SY: 1000;** MMY: 976

Volleyball
MY: 985, down 10 points over last year; SY: 957; MMY: 970

Multi-year APR Scores for Women's Sports

	WVB	WTEN	WGO	WSO	WBB
2009–2010	985	977	991	990	985
2008–2009	995	964	992	997	985
2007–2008	995	979	993	993	990
2006–2007	978	941	974	993	985
2005–2006	971	930	975	990	987

* Best MY Score in Metro Sports Conference
** Best SY Score in Metro Sports Conference

not convinced that some of the coaches understand these sheets. I'm sure the athletic director would appreciate some help communicating this information." Study which aspects of the design are effective and which can be improved. Consider the graphics, the text, and the overall design of each document. Then revise the documents so that they work together to better communicate the information.

DOCUMENT 1.5

APR Statistics for Men's Sports, 2009–2010

Bayside State

Department of Intercollegiate Athletics   7463 University Drive, Bayside, California 95524-1025

phone (707) 555-9175
fax (707) 555-1633
www.gullsports.com

Most Recent Multi-year and Single-year APR Scores for Men's Sports

Tennis
MY: 974, down 8 points over last year; SY: 967; MMY: 975

Golf
MY: 973, up 2 points over last year; SY: 1000;** MMY: 979

Soccer
MY: 951, up 32 points over last year; SY: 931; MMY: 967

Basketball
MY: 972, down 8 points over last year; SY: 964; MMY: 953

Water Polo
MY: 992,* up 5 points over last year; SY: 1000;** MMY: 977

Multi-year APR Scores for Men's Sports

	MTEN	MGO	MSO	MBB	MWP
2009–2010	974	973	951	972	992
2008–2009	982	971	919	950	987
2007–2008	980	967	914	989	971
2006–2007	980	983	900	928	979
2005–2006	974	977	895	893	970

* Best MY Score in Metro Sports Conference
** Best SY Score in Metro Sports Conference

DOCUMENT 1.6

APR Snapshot

Bayside State

Department of Intercollegiate Athletics   7463 University Drive, Bayside, California 95524-1025

phone (707) 555-9175
fax (707) 555-1633
www.gullsports.com

Academic Progress Rate (APR)
Snapshot

General Notes: The latest four-year Division I APR is 970, up 3 points over last year. The average four-year rate also rose in the high-profile sports of men's basketball (945, up 5), football (946, up 2), and baseball (959, up 5).

MY=Multi-year score; rates are based on the past four years' performance; statistical analysis includes 2006–2007 through 2009–2010 academic years

SY=Single-year score for 2009–2010 academic year.

MMY=Metro Conference Multi-year Average for sport for 2009–2010 academic year supplied by conference office.

Sport	Metro Multi-year Average for Sport
WVB	970
WTEN	976
WGO	981
WSO	978
WBB	963
MTEN	975
MGO	979
MSO	967
MBB	953
MWP	977

Calculations:

For each student-athlete (SA) on scholarship, 2 points each term/4 points each year:

Fall eligible=1 point; fall retention=1 point (thus, 2 points possible for fall term)

Spring eligible=1 point; Spring retention=1 point; (thus, 2 points possible for Spring term)

Retention defined: Is the SA returning to a full-time program of study next academic term? Yes=1 point

Eligibility defined: Would this student be eligible next term at this institution? Yes=1 point

MY calculation: Total points earned by all scholarship SAs on team for past four years, divided by total possible points, and then multiplied by 1,000.

SY Calculation: Total points earned by all scholarship SAs on team for past year, divided by total possible points, and then multiplied by 1,000

Task 3

● ● ● ●

Create Effective Graphics and Speaker's Notes

"Given some of the high-profile scandals at other institutions and recent news reports critical of intercollegiate sports in general, the faculty has become increasingly interested in how our student-athletes are performing in the classroom," Cheryl explains. "I'm going to address our institution's faculty senate next month. I have our most recent APR scores to illustrate the academic achievement of our student-athletes, but if I start going over a pile of statistics and using a bunch of acronyms, their eyes will glaze over and we'll likely not make a good impression. I think the scores provide strong evidence that the entire Department of Athletics makes academics a priority. Having the support of our faculty is critical to our continued success, both in the classroom and on the field or court. Can you help me create some graphics that effectively communicate the academic success of our student-athletes?"

Cheryl would like to make the following two arguments in her appearance before the faculty senate: (1) based on APR scores, many Bayside student-athletes are above average academically compared to other Division I schools, and (2) based on the most recent APR scores, Bayside teams do better academically than many teams in their conference.

Cheryl hands you the APR statistics for both women's sports (**Document 1.4**, p. 6) and men's sports (**Document 1.5**, p. 7) in 2009–2010, the APR scores for conference schools in 2009–2010 (**Document 1.6**, p. 8), and the APR scores for Division I schools in 2009–2010 (**Document 1.7**) and asks you to create one or more graphics for each argument. "It would be a huge help," she adds, "if you also suggested some speaking notes that I could use to introduce and explain each graphic to the faculty senate." Based on the two points Cheryl wants to make, determine the data and type of graphics that will best help her persuade her audience. Then write speaker's notes that will help Cheryl explain the graphics to her audience. When writing the speaker's notes, consider whether Cheryl will need to provide some background information to senate members before they can understand the graphics of the institution's APR scores.

DOCUMENT 1.7

Average Multi-year APR
Scores, 2009–2010

Bayside State

Department of Intercollegiate Athletics   7463 University Drive, Bayside, California 95524-1025

phone (707) 555-9175
fax (707) 555-1633
www.gullsports.com

Sport	All Division 1	Public Institutions	Private Institutions
WVB	978	975	986
WTEN	979	976	983
WGO	983	981	989
WSO	978	973	987
WBB	968	963	978
MTEN	970	966	978
MGO	971	967	979
MSO	967	960	974
MBB	945	937	961
MWP	972	977	968

Note: All information contained in this table is for four academic years. Based on NCAA Division I Academic Progress Rate (APR) data submitted for the 2006–2007, 2007–2008, 2008–2009, and 2009–2010 academic years.

Task 4

• • • •

Write a Press Release, and Integrate Text and Graphics

Each year, after APR scores are ready, the athletic director posts a press release on the department's Web site highlighting the academic success of Bayside teams compared to other teams in the Metro Conference. This year, she asks for your help in creating the press release. "I want to call attention to the teams that are outperforming their conference foes in the classroom," Cheryl explains. "I also want to help our fans understand how we are measuring academic success. Most people understand grade point averages, or GPAs. But few people outside of athletics understand APR scores." She hands you an outline for a press release (**Document 1.8**) that is missing the important data for this year. "This is just a start. Because this will be posted on our Web site, you may use color graphics, but keep the paragraphs short for easy reading. Keep the release under 750 words, and include the Bayside logo and at least one other graphic." Complete the press release, revising and reorganizing as necessary. You will need to consult **Documents 1.4** (p. 6) and **1.5** (p. 7) to gather any missing data. For more information on the development of APR scores, you may want to visit www.ncaa.org/wps/wcm/connect/public/ncaa/academics and read "How Academic Reform Is Measured."

DOCUMENT 1.8

Press Release Outline

Bayside State

Department of Intercollegiate Athletics   7463 University Drive, Bayside, California 95524-1025

phone (707) 555-9175
fax (707) 555-1633
www.gullsports.com

Courtesy: Bayside State College Sports Information

Release: date

Bayside, California—For the second straight year, the Bayside State College women's golf and soccer as well as men's water polo teams have the top multi-year academic performance rates (APR) in the Metro Sports Conference (MSC).

The NCAA announced its multi-year APR rates today (May 24). This statistical analysis covers . . . [describe what the multi-year APR covers].

Along with Bayside State leading the MSC's multi-year APR efforts in women's golf and soccer as well as men's water polo, the Gulls ranked second among conference schools in two other sports: women's basketball and men's basketball.

[mention this award in the release:] Last week the MSC honored the Gull women's golf and soccer teams, with an APR public recognition award. The award is presented to a school that posts multi-year APR rates in the top 10 percent for its specific sport.

[talk about our teams who posted a perfect *single-year* score of 1,000 points this past year] In the most recent single reporting year (2009–2010), the Gulls . . .

[also talk about how we compare to Division 1 teams and public institutions]

[We would like to include these quotes at appropriate places in the release. Please feel free to shorten, if necessary.]

"Once again Bayside State's excellent performance in the APR shows that our coaching and academic staffs are doing an outstanding job with our student-athletes. Academics is a core value of our department, and our student-athletes work hard every day to excel in the classroom."

Bayside State Director of Athletics Cheryl Mueller.

CONTINUED ➜

DOCUMENT 1.8

(continued)

"We expect our student-athletes in all sports to be students first and athletes second. The most recent APR rankings provide evidence that our students strive for the highest possible level of achievement. I am proud to see our teams ranked among the very best in the nation in both academics and athletics. Congratulations to our student-athletes, our coaches, and faculty for a job well done."

College President Dr. Catherine Asbury

"We take great pride in how our students perform in the classroom. The time we spend working with them on their academic progress really shows in these rankings. Success in the classroom has always been our top goal each year. When we succeed in the classroom and on the field, we know we have a successful overall program."

Gull Head Soccer Coach Peyton Brandt

"What a pleasure to work with such a dedicated group of student-athletes. Many of our fans don't understand how many hours these students devote to their studies."

Associate Athletic Director of Academic Services Madison Larson

[Include near the end of the release who developed the APR, what it is used for, and how it is calculated.]

The APR was developed by the NCAA in 2004 to measure the academic progress and performance for athletic programs at its member institutions . . . .

The APR is determined by using the eligibility and retention for each student-athlete on scholarship during a particular academic year . . . .

The APR is calculated . . .

Definitions and Descriptions
Helping Your Readers Understand

The Situation

When Mary Ann Petit, a neighbor of yours and the founder of Bayside United Green Squad (BUGS), learned of your technical communication background, she asked you to help with her community group. BUGS is a nonprofit volunteer group that offers environmental programs designed to teach young people ages 10 to 16 about the connections between environmental issues and food. Using a community garden (**Figure 2.1**) as a learning lab, BUGS volunteers provide local youth with hands-on science education on environmental issues affecting the local food they eat. To better reach their target audience, the board of directors has decided to enhance the BUGS Web site by including a blog.

Mary Ann explains that the success of BUGS has created not only a desire among program graduates to stay in touch, but also a demand for information on how young people can grow, harvest, prepare, and eat vegetables from their own backyard gardens. To address these demands, Mary Ann would like you to help start a blog to facilitate communication among young gardeners. By providing a mechanism for instructors and program graduates to comment on posts, Mary Ann hopes to build an online community.

Figure 2.1
Bayside's Community Garden

Although another volunteer will provide the technical skills to design the blog, Jenny Ukaegbu, the BUGS blogger, asks for your help writing some of the content for the initial blog posts, especially definitions and descriptions of gardening and blogging terms to help the group members understand how to effectively use and contribute to the blog. "In our first couple of blog posts," Jenny explains, "we need to teach the young gardeners a little about blogging and get them excited about reading our blog and posting comments—and we also need to start teaching them about backyard gardening."

Depending on the context, the blog posts will use parenthetical and sentence definitions as well as descriptions to help youth understand and succeed at blogging and gardening. Parenthetical definitions are placed directly in the text. When a simple word or phrase is not enough, the blog posts will feature pop-up sentence definitions for more formal clarification. Jenny shows you an example of two parenthetical definitions, one for *post* and one for *blogroll*, as well as a sentence definition for *phytoremediation* (**Figure 2.2**).

Figure 2.2
Model Parenthetical and Sentence Definitions

Parenthetical Definitions
With summer's growing season approaching, how about finding creative ways to enjoy your backyard garden? In this post (an individual entry in a blog), I provide tips for fun summer garden activities. For more ideas, be sure to check our blogroll, a list of links to other blogs, on the right sidebar.

The definition for *post* is enclosed in parentheses, and the definition for *blogroll* is enclosed in commas. Parenthetical definitions can also be introduced with a colon or a dash.

Sentence Definition
Phytoremediation is a <u>cleanup technology</u> using <u>living plants</u> to remove or detoxify environmental contaminants from soil and water.

The item to be defined (*phytoremediation*) is placed in a <u>group of similar items</u> and then distinguished from them by stating the <u>key distinction</u> between the item being defined and the other items in the group.

The Challenge

Being able to learn new material quickly is an important job skill for technical communicators, who often work on new product lines or on projects involving the contributions of a variety of experts. Like many nonprofits, BUGS recruits volunteers with a variety of interests, including people with writing and technology skills to support the educational programs. Most BUGS volunteers commit to working at least 10 hours per month as program instructors or in the BUGS garden. As the first volunteer with a strong background in writing, you will spend most of your time helping improve the program materials and writing content for the Web site. Because you might not have much experience with gardening or blogging, your challenge is to research and learn new terminology and then clearly explain these terms to the young gardeners.

Your Job

To download electronic copies of documents in this chapter, click on **Case 2 Documents** on **bedfordstmartins.com /techdocs**.

The success of this new interactive blog depends on the young gardeners' understanding the definitions and descriptions. If they cannot understand a gardening term or a blog feature, they may give up on their backyard gardens or fail to make use of the blog. Your job is to analyze the site's audience and purpose and then help the BUGS blogger explain backyard gardening and blogs. To prepare the blog for its launch, you may be asked to do the following:

▶ Analyze parenthetical and sentence definitions.

▶ Evaluate and revise descriptions.

▶ Identify where a graphic would help clarify and complement a written definition or description.

▶ Write several definitions and descriptions for gardening terms that young people might not know.

Your instructor will tell you which of the tasks you are to complete.

Get started on your job.

When You're Finished

Reflecting on This Case In a 250- to 500-word response to your instructor, discuss (a) what you learned from this case, (b) how you could relate this case to work situations you will face in your chosen career, and, if applicable, (c) the ways in which this case compares to similar situations you have already faced at work. Your instructor will tell you whether your response should be submitted as a memo, an e-mail, or a journal entry, or in a different format.

Moving beyond This Case Write a 500- to 1,000-word description of a piece of equipment or a process used in your field. Include appropriate graphics. On a separate sheet, briefly describe your audience and the purpose of the document in which your description would likely appear.

Task 1

● ● ● ●

Analyze Parenthetical and Sentence Definitions

During a call, Jenny tells you, "Yesterday, I brainstormed a list of terms I think I will use in my early blog posts. I just sent you two e-mails with several definitions I wrote (**Documents 2.3** and **2.4**). I would like your feedback on whether the definitions are effectively written and appropriate for our BUGS audience before I use them in a post." She adds, "Don't be afraid to use the Internet to learn about any unfamiliar terms." She then asks you to write an e-mail in which you identify by letter the effective definitions, include revised versions of those that are flawed, and identify any terms that are not appropriate for the blog's intended audience.

DOCUMENT 2.3

Parenthetical Definitions

To: [your name]
Subject: Definitions for your review

I'd like my first post to focus on "how to use this blog." I'd like you to review these parenthetical definitions before I use them and let me know which are OK as written and which need work. For those definitions needing work, please include a revised version in your response. Also, if you think I've included a term that is too technical for our audience, let me know.

Thanks,
Jenny

Parenthetical Definitions

A. When people started sharing personal stories and comments on the Internet in the 1990s, these writings were called a Weblog (later just called a blog). Soon people could access the blogosphere (the Internet community of over 150 million public blogs) to read bloggers' observations on almost any topic of interest.

B. To understand what topics are posted most frequently, our tag cloud, a visual cloud of tags emphasizing the most frequent tags appearing in our blog, is a good resource. Located on the right sidebar and organized alphabetically, the bigger the word appears the more frequent the topic appears in our blog.

C. You'll note on our index page (or index.html, a traditional file name for the page you are on currently) we provide a brief overview of BUGS and explain the purpose of our blog.

D. Blogopotamus: Not an alien creature but a very long, bloated blog post.

E. A form of audio broadcasting on the Internet typically using an MP3 format, podcasting, is like downloading and listening to music. However, rather than music, you listen to someone talk about a topic of interest such as organic gardening.

DOCUMENT 2.4

Sentence Definitions

To: [your name]
Subject: Definitions for your review

I just wanted you to look over some of these sentence definitions, too. I've included both blogging and gardening terms. I plan to link these pop-up definitions from a glossary page. Let me know which are OK as written and which need work. For those definitions needing work, please include a revised version in your response.

Thanks,
Jenny

Blogging Definitions

A. Tags are, as the name suggests, tags used to categorize, label, and organize lots of blog content.

B. The archive, usually categorized by month, is where you will find a collection of previous posts to a blog.

C. Less susceptible to link rot, a permalink is a permanent link to a specific blog post even after it has been archived.

D. Plug-ins are a type of small computer code that add functionality and new features to a blog such as social networking plug-ins.

E. Rather than having to visit many blogs to check if a new post is available, a Web feed allows you to get regular updates from one spot when content is updated.

Gardening Definitions

F. Bedding plants (mainly quick and colorful annuals) are nursery grown and used for growing in beds.

G. A dibble stick is used by gardeners to makes holes in the soil for seeds or young plants.

H. A taproot (not all plants have a taproot) is a thick root (the part below the surface used to secure and nourish) that grows directly down (in most cases).

I. Scarification is what is performed when a seed's shell is scratched or nicked in order to facilitate germination.

J. Hardpan: the impervious layer of clay (a naturally occurring aluminum silicate) lying beneath the top soil—one of the most difficult types of soil a gardener can have the misfortune to encounter in his or her backyard.

Task 2

• • • •

Evaluate and Revise Descriptions

"I think some of the longer descriptions I plan to include miss their mark," Jenny tells you, "but I'm unsure how to revise them. Could you take a look at four passages I'm having trouble with, identify the techniques I use, and then evaluate the effectiveness of these techniques? Please include revised passages." She says she will attach the passages and her annotations to an e-mail message (**Documents 2.5** to **2.8**), and she would like you to respond by e-mail. When you evaluate the passages, Jenny wants you to consider carefully the blog's audience and the organization's purpose for posting the information on the blog.

DOCUMENT 2.5

Object Description of
Beneficial Insects

> What tone do you think is appropriate for our blog posts?

Everyone considers a bug to be a pest, right? Not so! There are actually insects that are beneficial to our environment—especially in the garden! Just what is a beneficial insect? A beneficial insect is an organic, natural pest control method that also has the ability to pollinate. They can reduce and even eliminate the need to use chemical pesticides that can be harmful and may not even work. Because they are organic, they are not harmful to the plants, people, or pets. Now keep in mind, they are not a miracle cure for all your garden woes, but you won't encounter pesticide resistant pests if you utilize beneficial insects because there is no immunity to being eaten!

So now what? Which insects are considered beneficial and what do they do? Below is a list of several types of beneficial insects that are safe, organic, and will do the work for you.

Ladybugs (see pic): preys on aphids and mites; Minute Pirate Bug: preys on spider mites; Big Eye Bug: predator to mites; Assassin Bug: predator; Damsel Bug: preys on cabbage worms and aphids;

Soldier Beetle: consumes aphids, grasshopper eggs, caterpillars, also a minor pollinator;

Honey and Predator Bees: facilitate pollination, propagation, and fruit production, kill pests;

Trichogramma Wasp: parasitizes the eggs of agricultural pest insects and easy to release in fields

Others include spiders, preying mantis, scarab beetle and many more! Where can we find the beneficial insects for our personal use in the garden? Many occur naturally and we kill them ourselves because we aren't aware of their capabilities and many are just plain frightened of them. Several are available for purchase at the local garden store or on the Internet. Search the keywords 'beneficial insect', identify what will work in your garden and make the purchase and let them get to work. Remember, it's a bug eat bug world out there!

DOCUMENT 2.6

Mechanism Description of Garden Tiller

This description starts with an informal tone, then switches to a more formal tone.

Do you think that this historical information is important?

Picture yourself in ancient Roman times cultivating the soil yourself with hand tools and maybe even a plow and horse. Now picture yourself today doing the same manual labor in the garden. Yuck! Why not purchase a garden tiller and let it do the heavy work for you!

Just what is a garden tiller? A garden tiller is a motorized rotary plow that 'tills' or cultivates the soil. Functioning independently (being operated by a human), or attached to a tractor (also operated by a human), the rotary blades continuously turn the soil providing easier access to nutrients and improving the soil structure for planting.

The garden tiller was in its initial stages of creation in 1912 by Arthur Clifford in Australia. By 1922, he found that it worked well and he established the Howard Rotavator Company, Ltd. The Rotovator is still in use today and is self propelled, moving forward and backwards. Dr. Konrad von Meyenburg also created a "machine for mechanical tilling" in 1910 and Troy-Bilt followed much later with the 'Rototiller' in the 1930's. Nowadays there are literally hundreds of garden tillers available at your fingertips. Depending on the needs of your garden, how do you choose what will work? Here are a few simple hints to choosing a garden tiller.

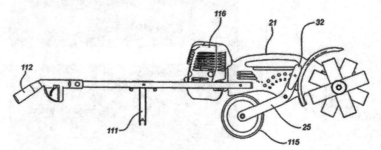

1. Choose something easy to move and lightweight.
2. Choose either electric or gas powered.
3. Choose front or rear tined; front maintains already cultivated soil whereas rear is the workhorse of the garden for uncultivated soil.
4. Check the horsepower: increased HP for breaking up new ground and decreased HP for the casual garden.
5. Check the prices! There are lots of brand names and prices available. Consider what works best for you at the price you can afford.

Get to the garden and cultivate!

DOCUMENT 2.7

Process Description of
Erosion Control

Do you think these
questions help to frame the
discussion that follows?

So you have a garden and you want to prevent a common occurrence called erosion control. What is it? How do you prevent such a thing if you don't really know what it is?

Erosion control is the practice of preventing or controlling wind or water erosion in agriculture—for our purposes—gardening. This, in turn, is important to prevent water pollution and soil loss. Erosion control often involves the use of a physical barrier to absorb some of the energy from the force causing the erosion. Obviously we all can't have an indoor greenhouse to protect from the wind or a barreling rainstorm, so you can utilize other forms of barriers, such as vegetation or rock. Fred Turner, master gardener, uses a raised platform bed (below) and rock barriers in the garden area.

Should we include
this mention of a local
gardener's erosion control
method?

I'm undecided whether
I like the more technical
parts of this passage (e.g.,
cellular confinement and
bioswales). I'm wondering
if our readers would be
interested in such detail.
What do you think?

There are many different items that may be used to prevent erosion control. Mulching is a great use of organic material that can cover the garden to protect it from erosion and provide nutrient as well. Perennial plants can also be used to buffer the surrounding areas of the garden. Other examples include, cellular confinement systems, contour plowing, bioswales, crop rotation, and reforestation. Not all of these would be prudent in the average home garden, but would be used in agriculture and urban landscaping.

So when you consider creating your garden and landscaping around it, consider what you need to prevent erosion control. Then, not only will the yard look good, you will have a thriving and fruitful harvest from the garden.

DOCUMENT 2.8

Object Description of
USDA Plant Hardiness
Zone 8b

The USDA Hardiness Zone map divides North America into 11 separates zones. Each zone is 10 degrees F warmer (or colder) in an average winter than the adjacent zone. Some of the zones are further divided into 'a' and 'b' categories—depending on the reference. Zone 8b skirts near the western and southern borders of the United States where the average climate has warm winters and a long growing season. The minimum temperature for Zone 8 ranges from 10-20 degrees F (-12.3 to -6.6 Celsius) and Zone 8b further delineates that range from 15-20 degrees F (-9.4 to -6.7 Celsius). Geographically speaking, certain plants are capable of growing better in certain climates and the Zone map strives to give the unknowing gardener a better idea of what will or will not grow, depending on the zone map. The USDA map is most esteemed in the eastern half of the country where the land is relatively flat and the zoning works well. There are, however, drawbacks to using the hardiness zones in the west. The map does not take into consideration summer heat levels or amount of snow cover in the winter. Weather coming from the Pacific Ocean becomes less humid as it crosses the mountain ranges in the west and becomes much dryer, so the climate in Seattle is very different than the climate in Bayside, but both are in Zone 8b.

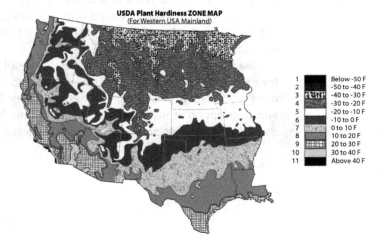

USDA Plant Hardiness ZONE MAP
(For Western USA Mainland)

1	Below -50 F
2	-50 to -40 F
3	-40 to -30 F
4	-30 to -20 F
5	-20 to -10 F
6	-10 to 0 F
7	0 to 10 F
8	10 to 20 F
9	20 to 30 F
10	30 to 40 F
11	Above 40 F

Do you think we should
even include info on the
Latin names for plants?

So, what can you plant in Zone 8b? Lots of things! The Arbutus Unedo (strawberry tree), Choisya terhata (Mexican orange), and Olearia x haastii (New Zealand daisy bush) to name a few. If you have no idea of what those plants are, stick with the more common vegetables such as spinach, kale, onions, celery, broccoli and cauliflower. They are planted at different times of the year, so reference your favorite garden site for that information. Common fruits for Zone 8b include blueberry, raspberry, kiwi, grapes, and apples. You can even plant nuts in Zone 8b as well. Good choices would be almonds, pine nuts and pistachios. In the future, scientists and gardeners are working on revising the USDA Hardiness Zone Map to include recent changes in the climate. This may be even more helpful than the current map. For now, living in Zone 8b, creating a super fantastic backyard Bayside garden can yield some pretty tasty eats!

Task 3

• • • •

Clarify Definitions with Graphics

While discussing some of the organization's printed educational materials and planned blog posts with you, Jenny remarks, "I'm a little concerned about how wordy our blog might appear. Our young gardeners just won't read dense text. I'm sure we could add some visual interest and reduce the text-heavy feel, but I'm just not sure where to do it."

Because BUGS participants are just beginning to learn gardening terminology, Jenny explains that she would like to use graphics as well as text in some of the posts. She asks you to review **Documents 2.5** to **2.8** (pp. 20–23) and evaluate whether the proposed graphics would clarify and complement the text. If you decide that the proposed graphics are not suitable, or you determine that additional graphics would help, suggest alternatives. In an e-mail to Jenny, describe how one or more graphics would help, and describe how to best integrate the graphics with the text. Remember that these graphics will be displayed online in a blog post.

Finally, Jenny would like to convert some of her blog posts to handouts for use with various BUGS programs. Select one of the descriptions in **Documents 2.5** to **2.8**, and design a one-page color handout suitable for printing and distribution at BUGS educational programs. You will likely need to further develop the description and locate suitable graphics.

Task 4

• • • •

Write Definitions

A BUGS instructor, Terra Hughes, has identified several objects, mechanisms, and processes that young gardeners have asked to be defined. She has sent you an e-mail (**Document 2.9**) asking you to write several sentence definitions and a few extended descriptions.

DOCUMENT 2.9

E-mail with Gardening Terms to Be Defined

To: [your name]
Subject: Gardening terms to be defined for our site

I've been keeping track of the terms that the gardeners enrolled in our programs frequently ask about. I'll be asking several instructors to help me define these terms. However, I'm giving you first pick of the terms you would like to work on. Choose three terms from the first category and one term from each of the remaining categories. Let me know which ones you pick, and then send me your definitions in an e-mail.

Thanks,
Terra

Category 1 (sentence definitions):
   A. Alkaline
   B. Annual
   C. Perennial
   D. Seedling

Category 2 (30- to 50-word description of an object):
   A. Ground cover
   B. Capillary matting
   C. Compost
   D. Fertilizer

Category 3 (50- to 100-word description of a mechanism):
   A. Fertilizer injector
   B. Seed spreader
   C. Drip irrigation
   D. Water dechlorinator

Category 4 (100- to 300-word description of a process):
   A. Photosynthesis
   B. Dividing
   C. Evaporation
   D. Leaching

Correspondence
Considering Your Reader's Point of View

The Situation

Axel Geirsson (**Figure 3.1**), an accomplished adventurer who has twice reached the summit of Mount Everest, recently moved to the United States and opened a small climbing and outdoor specialty shop in Boulder, Colorado. Onward & Upward Outfitters is focused on providing climbing enthusiasts with high-quality outdoor gear and apparel. It boasts an extensive selection of climbing hardware, harnesses, ropes, slings, helmets, and footwear. The store is staffed by an extremely knowledgeable team, yet Axel has noticed that his staff has some difficulty explaining technical concepts to customers who do not have extensive outdoor or climbing experience. "My staff from time to time gets caught up in the enthusiasm of helping customers prepare for outdoor adventures and

assumes that customers already know the basics when, in fact, many do not," Axel says. Consequently, some customers leave the store believing they know how to use their new equipment properly, only to be stranded miles from a mobile-phone signal, sometimes in the middle of the night or during a thunderstorm.

The staff's occasional misjudgment of their customers' background knowledge has generated a few complaints. The staff can successfully address the concerns of those customers who return to the store or call to report a problem. However, customers who e-mail or send a letter do not always get an effective response. Some disgruntled customers have begun posting critical comments on popular outdoor blogs and on microblogging sites like Facebook and Twitter. Rather than resolving customer problems,

Figure 3.1
Axel Geirsson

the store's correspondence merely generates more problems. As a result, Axel has hired a public relations firm, which has assigned you, a communications consultant at the firm, to help Axel improve his store's correspondence.

The Challenge

This case requires that you put yourself in the reader's shoes. Consider how a reader would respond to the content, design, and tone of the correspondence. Likewise, this case requires you to think about the purpose of each piece of correspondence: what does the writer want his or her reader to know, do, or believe as a result of reading the response? You must then find the best strategy for achieving the writer's purpose.

Your Job

To download electronic copies of documents in this chapter, click on **Case 3 Documents** on bedfordstmartins.com /techdocs.

Although the staff at Onward & Upward Outfitters are expert climbers, they need your assistance improving their company's correspondence so they can provide better customer service. You are asked to use your background in communication to analyze correspondence surrounding a particular rain gear product sold by Onward & Upward Outfitters. You may be asked to do the following:

▶ Evaluate and revise the company's internal memo.

▶ Examine a customer's claim letter, and revise the corresponding adjustment letter.

▶ Respond to two claim letters.

▶ Write one to three microblog posts and a blog post addressing a common customer complaint.

Your instructor will tell you which of the tasks you are to complete.

Get started on your job.

When You're Finished

Reflecting on This Case   In a 250- to 500-word response to your instructor, discuss (a) what you learned from this case, (b) how you could relate this case to work situations you will face in your chosen career, and, if applicable, (c) the ways in which this case compares to similar situations you have already faced at work. Your instructor will tell you whether your response should be submitted as a memo, an e-mail, or a journal entry, or in a different format.

Moving beyond This Case   In small groups, brainstorm a list of instances in which group members purchased a defective product or received poor service. Then choose one of these instances and work together to write either a polite, reasonable claim letter or a blog post designed to attract the attention of a company representative and get the problem resolved.

Task 1

● ● ● ● ●

Evaluate and Revise an Internal Memo

A few months ago, Onward & Upward Outfitters began selling lightweight rain gear featuring a new thin, waterproof, breathable layer of fabric called H_2O Pro-TechT. A few customers have complained that the rain gear seems to leak. "I really don't think the fabric is defective. It's just that our customers are not being properly educated about what they can expect from H_2O Pro-TechT or how to care for the fabric," Axel explains. He has addressed the situation by drafting a memo to store personnel (**Document 3.2**) stating the store's policy regarding H_2O Pro-TechT apparel. Evaluate the effectiveness of Axel's draft, and then revise it. Write Axel an e-mail in which you justify your revisions, and attach your revised memo for his review. Axel also encourages you to improve the design of his draft.

DOCUMENT 3.2

Memo on H_2O Pro-TechT Features

MEMO

To: Onward & Upward Staff
From: Axel
Date: May 25, 2012

As you know, we've been selling a lot of the H_2O Pro-TechT product line. Unfortunately, we've also been getting customer complaints that the H_2O Pro-TechT fabric leaks, leaves them damp, or leaves them cold. I don't think the fabric is defective. The fabric is highly tested and undergoes strenuous quality checks. You're not effectively educating the customers about what they can and cannot expect from H_2O Pro-TechT.

H_2O Pro-TechT is not a space suit. Water can and will get in. When discussing H_2O Pro-TechT rain gear, remember to emphasize the following: most feelings of dampness are not the result of leaks but of sweat. Under exertion, humans produce a surprising amount of sweat. Customers can address this problem by using the ventilation zippers (i.e., pit zips), wearing moisture-wicking clothes underneath, and layering clothes.

When handling customer concerns, try to find out if the product is defective or if the customer did something to reduce the product's performance. For example, did they wait until they were already wet before putting on the rain gear? Did they have the pit zips open and allow water in? What type of clothing were they wearing underneath? Cotton clothes could leave them feeling damp during exertion. Too much clothing might lead to overheating, sweating, and an impression that the gear leaks. How long have they had the gear, and have they washed it recently? Dirt, crud, and salt water reduce the effectiveness of H_2O Pro-TechT. Sometimes a good washing will dramatically improve the performance of the gear.

Bottom line: If customers are concerned, they can bring the gear back and we'll take a look at it. If the seams or other material look defective, we'll exchange it for another. This is a good product—perhaps the best on the market. By taking just a few minutes to talk with customers, we can help them avoid common mistakes and get the most out of their gear.

Onward and upward!

Axel

Task 2

• • • •

Evaluate and Revise a Response to a Customer

Onward & Upward Outfitters recently received a complaint (**Document 3.3**) from a customer who had purchased H₂O Pro-TechT rain gear. The sales specialist who originally helped the customer volunteered to write a response (**Document 3.4**). Before mailing the response, Axel has asked you to examine the correspondence and report to him on the effectiveness of his staff member's response. Study the claim and adjustment letters. Does the claim letter seem polite, reasonable, and specific? Is the response fair and reasonable? Revise the response letter. Write Axel an e-mail in which you justify your revisions, and attach your revised response for his review.

DOCUMENT 3.3

Customer Claim Letter

5000 Wilderness PL
Boulder, CO 80301
May 17, 2012

Mr. Axel Geirsson
Onward & Upward Outfitters
701 Butte St
Boulder, CO 80301

Dear Mr. Geirsson:

On a recommendation from one of your sales specialists, Brook Rodney, I recently purchased a H2O Pro-Techt rain jacket (women's medium). Brook told me that the H2O Pro-Techt was not only lightweight and packable but also waterproof.

Last weekend I accompanied my son's Boy Scout troop on an overnight camping trip. As you probably know, it rained most of the weekend. When the dark clouds began to roll in while we were making camp, I didn't much worry because I thought I had a waterproof rain jacket. I was wrong. The jacket just didn't keep me very warm and dry. While the scouts completed their compass skills and other activities, I spent the entire weekend damp and cold. My son's inexpensive PVC poncho kept him drier than my expensive "waterproof" jacket.

I feel that Brook misrepresented the features of the rain jacket—especially the waterproof feature. As a store advertising itself as a "place where the experts gear up," I expected a much higher quality product. I would like a full refund of $157.43 (attached is a copy of the receipt) for the jacket.

Before my family or the scout troop shops again at Onward & Upward Outfitters, I need to be convinced that your products will hold up during our outdoor activities, including those in the rain.

Sincerely,

Elise Smith

Elise Smith

DOCUMENT 3.4

Response to Customer
Claim Letter

www.onwardandupward.com
701 Butte St
Boulder, CO 80301
(303) 555-6464

May 21, 2012

Elise Smith
5000 Wilderness PL
Boulder, CO 80301

Elise,

Referring to your letter dated May 17, I did not misrepresent the H_2O Pro-TechT rain jacket. H_2O Pro-TechT is the best breathable, waterproof rain gear available. I suspect you did not use the jacket properly.

H_2O Pro-TechT, like all other waterproof/breathable fabrics, is not 100 percent waterproof and flawlessly breathable. However, you cannot find a better combination of water repellency and vapor transmission (sweat). Yes, PVC ponchos are inexpensive and certainly waterproof. However, they are heavy and bulky, not to mention extremely uncomfortable under just moderate exertion or outdoor temperature.

You mentioned that you spent the weekend cold and damp. First, you need to realize that H_2O Pro-TechT is not a good insulator from the cold (wind yes, cold temperatures, no). It may be that you just need to wear a heavier, moisture-wicking layer underneath. On the other hand, you may have been wearing too much or a cotton fabric (remember, cotton kills) under the jacket. The dampness you felt may have been the sweat you built up while making camp. If this is the case, remember to use the ventilation zippers (pit zips) to allow your body heat and sweat to escape. You can check to see if your jacket does indeed leak by wearing it and standing in your shower or lawn sprinklers.

See you up there,

Brook Rodney

Brook Rodney
Climbing Sales Specialist

Task 3

• • ● •

Respond to Customer Claims

Axel received two e-mails over a holiday weekend regarding H$_2$O Pro-TechT apparel sold at his store. Write an e-mail denying the request made in the first message (**Document 3.5**), and write another granting the request in the second message (**Document 3.6**).

DOCUMENT 3.5

Customer Claim E-mail 1

To: Onward & Upward Outfitters
Subject: Raining Chickens and Ducks

It's called foul (fowl) weather, and that's what we were caught in this week-end. That breathable rain gear you sold us last week failed us. Sure, H20 PRO — WHATEVER works if you're standing still not doing any physical exercise, but your claim that the rain gear we purchased was breathable enough to keep us dry from sweat while we're actually hiking in three-season conditions is just FALSE. There was an upside, however. At least the rain gear helped retain our body heat, even if we were drenched with sweat instead of rain, so being wet and warm is better than being wet and cold. IMHO your rain gear just doesn't work. While it was raining, we found a better solution: it's just better to keep that "high tech" rain gear you sold us inside our packs and get drenched and just hike with a t-shirt. My wife just hiked in one of those sport shirt bra type things and was OK most of the time. Then when we got to camp we put that "high tech" rain gear on to retain some body heat. I know you won't give us our money back, so how about sending us some "high tech" t-shirts to wear when it rains?

Not holding our breath,
Rick & Kim Hart

DOCUMENT 3.6

Customer Claim E-mail 2

To: Onward & Upward Outfitters
Subject: H20 Pro-TechT Raingear's poor performance

Dear Onward & Upward Outfitters:

Since you opened your shop, I have purchased several pieces of outdoor equipment from you. I have always been impressed with your friendly and knowledgeable staff as well as your high-quality gear that doesn't let me down when I need it. Recently, I purchased the H2O Pro-TechT ultralight waterproof jacket (#568870) and pants (#651932).

This weekend I got a chance to "field test" my new rain gear. Unfortunately, I spent most of my trip wet and cold. My shirt was soaked only a few hours after I put on the jacket. I couldn't tell if the seams were leaking or if the jacket just didn't "work."

I know that H2O Pro-TechT is relatively new. Am I the only one who has had problems with the H2O Pro-TechT rain gear? As I trudged along the trail, I kept thinking maybe I was just unlucky and bought a defective set of rain gear. If I brought my rain gear back, would you be willing to take a look at them? And, if nothing seems wrong, may I return the rain gear? I don't want to spend another day in the rain like I did this weekend.

Sincerely,
Jeff Collins

Task 4

• • • •

Write Microblogs and a Blog Post

Axel shows you his internal memo on the H$_2$O Pro-TechT product line (**Document 3.2**, p. 29) as well as recent customer complaints (**Documents 3.3**, **3.5**, and **3.6**, pp. 30, 32, 33). He also shows you copies of a few microblog comments from customers (**Document 3.7**) and a blog post (**Document 3.8**) from a local climber critical of the store and the product. Axel sighs. Recently, the store has established an online presence with a blog and a microblogging site, and Axel would like your help addressing this customer issue using these Web 2.0 tools. "I'd like you to draft one to three microblog posts, no more than 140 characters each, and a longer blog post of about 300 to 500 words. Although I want these to respond specifically to the concerns raised about our H$_2$O Pro-TechT line of clothing, I also want to make a strong statement about our overall commitment to our customers. We need to build relationships with customers and establish a positive presence in the blogosphere."

DOCUMENT 3.7

Microblog Posts Critical of Company

Climber7890
#H20ProTechT jacket leaked. OnUpOutfitters not helpful. Way overpriced for performance.
1 hour ago

HighAlt4567
Onward & Upward staff don't know what they're talking about. I'm taking my business elsewhere.
3 hours ago

Mtn_Ace
Waterproof maybe, but #H20ProTechT breathable NOT. More of an urban raincoat around town, not for hiking. No point wearing if above 70 degrees.
20 hours ago

PeakClimber987
#H20ProTechT Xcellent outer shell for gen outdoor hiking, biking. Some climbers think it fragile but durable for most uses.
20 May

Summit5678
Wore Onward & Upward #H20ProTechT jacket on mildly wet morning and was immediately cold and soaked: colder than had I not had on the jacket.
5 May

On Belay Blog

Friday, May 18, 2012

Gear review of H20 Pro-TechT Jacket

OK, I've been wearing this all season, so it's time for a review. I've carried my H20 jacket with me every climbing trip, adding layers underneath depending on conditions. The H20 Pro-TechT waterproof jacket sold by Onward & Upward Outfitters comes with clean styling and multisport versatility. Here's the specs: nylon blend fabric, athletic fit, windproof, waterproof, breathable laminate (not), helmet-compatible hood, and thigh length for max coverage, laminated zippers, articulated elbows, convenient Napoleon and upper-arm pockets, and waist cord. Sounds great, right? Not so fast.

I made my almost weekly shopping trip to Onward & Upward Outfitters, looking for a minimalist alpine jacket for use while climbing. According to the sales staff at Onward & Upward Outfitters, "This jacket will keep you bone-dry in a hurricane." I get they are trying to sound enthusiastic and sell me the product, but they just don't know what they're talking about. The staff is a bunch of young college kids in sandals who have never actually been climbing or been caught in a lengthy rainstorm on the side of a mountain. Contrary to the sales pitch, the fabric is NOT breathable. Unless you plan to wear it just while you sit around camp, the jacket is just not suitable. I ended up a clammy mess on most of my climbs. With a layer underneath, I can avoid most of the moisture issues. This seems like a cheap fix for an expensive but flawed product.

Other concerns: That upper-arm pocket: worthless and a pain to open and close. The hood is not helmet-adjustable. The main zipper is not waterproof. The lining is starting to delaminate. So, all in all, a big thumbs down for the H20 Pro-TechT jacket and the guy who sold it to me.

I expect more from a jacket in this price range and question whether Onward & Upward should even be selling this product.

Share it

Share this on Facebook
Tweet this
View stats

Get more gadgets for your site

Share |

+1

Like us on Facebook!

Recent Posts

▼ 2012 (157)
  ▶ May (33)
  ▶ April (32)
  ▶ March (29)
  ▶ February (28)
  ▶ January (35)

▶ 2011 (329)
▶ 2010 (396)
▶ 2009 (466)
▶ 2008 (377)
▶ 2007 (71)
▶ 2000 (1)

DOCUMENT 3.8
Blog Post Critical of Company

Proposals
Seeing Proposals through Reviewers' Eyes

The Situation

The Heathcot-Ann Foundation is a charitable organization established in 2008 with assistance from the estate of Drs. Heathcot and Ann Turner. Its mission is to provide a pool of philanthropic capital to fund a variety of wellness and prevention programs in Idaho. In 2011, the foundation awarded a total of $14,000 to seven nonprofit projects in Boise, McCall, Caldwell, and Weiser. Foundation recipients in 2011 included community-sponsored clinics, a domestic violence crisis line, a school-based childhood obesity program, and a tobacco-cessation program for hospitalized patients.

As a new social media intern, you are responsible for developing content for the foundation's Web site, blog, microblog, and other social-networking sites. To familiarize yourself with the foundation's work, you have been asked by the board of directors to review proposals submitted to the foundation. The vast majority of the foundation's awards address four goals:

- To encourage innovative health and safety programs

- To improve the way health and safety services are provided to at-risk groups

- To foster collaboration among health and safety providers

- To support the teaching of healthy lifestyles

To help you evaluate each submission, the board of directors has given you the following evaluation criteria:

- *Need for the project.* Does the proposal describe specific gaps or weaknesses in health or safety services?

- *Goal.* Does the proposal clearly describe what the project will accomplish?

- *Project approach.* Does the proposal describe an effective method for accomplishing the goal?

- *Cost.* Can the project be accomplished with the proposed budget? Are all budget items justified in the proposal?

- *Impact.* If successful, who will benefit, and to what extent?

The board has also provided you with an example (**Figure 4.1**) of a winning proposal from last quarter. Included are brief comments from the reviewers.

"Unfortunately, we can't fund all the deserving proposals we receive," explains Jessica Bureau, the president of the foundation's board. "The recent economic

Figure 4.1
Sample Funded Proposal

Contact Person: Shirlee Williams
King County Senior Center Health Program
823 Pershing Drive
Boise, ID 83704

Project: Breast Self-Examination (BSE) Outreach Program

Problem Statement
According to the American Cncer Society, one in eight women will develop breast cancer in their lifetime and two-thirds of those women will be over 50. Although most breast lumps are discovered by women themselves, only 35% practice monthly breast self-exams (BSE). However, if the cancer is detected early (localized), the five-year survival rate is 96% compare to 75% if the cancer has spread regionally.

Therefore, there is a need to educate and encourage the high-risk group (women over 50) to perform monthly BSE. According to the Avon's Breast Cancer Awareness Crusade survey, more than half of the women surveyed said they would be more motivated to get mammograms and clinical breast exams if encouraged by people close to them and 40% name volunteers from health programs among the group of motivators.

Need for project: Although a spelling error in the first sentence concerned some reviewers, all agreed that the writer uses quantitative data (for example, numbers and percentages) from a credible source to establish the project's importance. The problem statement relates the project to foundation goals and uses survey data to suggest that the Senior Center is a strong candidate to conduct the project.

Goals
1. Educate the high-risk population in King County about breast cancer and train them to perform BSE.
2. Conduct twenty BSE workshops with at least twenty-five attendees per workshop.

Goals: Succinctly explains what the project will accomplish by identifying the target population, indicating the project's scope, and describing project events.

Approach
Our strategy for achieving our goals is the following:

- Publicize BSE workshops through civic organizations, church groups, retirement homes, senior centers, press releases, newsletters, and flyers.
- Hold BSE workshops in conjunction with Breast cancer Awareness Month (October).
- Use facilities at the Senior Center and volunteer health educators to teach workshops.

BSE workshops will cover the following topics:

- Why do BSE (statistics, benefits, excuses)?
- When to do BSE (body cycles & types, frequency of BSE)
- How to do BSE (breast model kits, diagrams, shower cards)

Approach: Approach is effective, covering publicity plans, the project's timing, and cost-savings activities. Also demonstrates how the project will educate the target population by describing the content of the BSE workshops.

Budget

Item	Notes	Cost
Publicity	Flyers, mailings, newspaper ads	$200
Workshop Materials	Breast model kits (4), education kits (2)	500
Participant Materials	500 handouts, shower cards, BSE pads	1,800
Refreshments	Soda @ $1 x 25 participants x 20 workshops	500
	Total:	$3,000

Cost: Uses a three-column table to clearly present a reasonably detailed budget. Dollar amounts are accurately totaled. Reviewers felt $500 was excessive for refreshments and not directly related to the project's goals. As a result, the project was awarded $2,500.

Impact
Achieve a higher percentage of early detection of breast cancer among participants in the King County high-risk group who perform BSE with a resulting higher survival rate due to earlier treatment and follow-up care.

Impact: Although wordy, this section clearly aligns the project with the foundation's goals by stating how a King County at-risk group will be taught a healthy lifestyle skill that will likely increase the group's survival rates.

downturn has resulted in a tenfold increase in requests for funding. However, the board has allocated no more than $3,000 for awards this quarter. We want you to recommend which projects should be funded. When making award decisions, you may decide to fund a project fully, partially, or not at all."

The Challenge

After proposals are submitted and distributed to reviewers, they must be quickly read and evaluated. Then a consensus must be reached on which proposals should be funded and which should be rejected. This can be a daunting process when dozens of proposals, each hundreds of pages in length, must be reviewed. Consequently, clear, concise, and easy-to-read proposals — in other words, *reviewer-friendly* proposals — improve an organization's chances of being funded. Your challenge, the same one all reviewers face, is to identify the proposals that have the best chances of furthering the *funding source's* goals.

Your Job

To download electronic copies of documents in this chapter, click on **Case 4 Documents** on bedfordstmartins.com /techdocs.

The Heathcot-Ann Foundation needs your help evaluating the most recent proposal submissions. You may be asked to do the following while working within the financial constraints outlined by the foundation:

▶ Evaluate proposals using the criteria established by the foundation, and recommend which, if any, projects should be funded.

▶ Reach a consensus with other reviewers, reflect on your review process, and provide online advice to proposal writers.

▶ Write e-mail messages to each organization whose proposal you recommended be rejected or only partially funded.

▶ Develop a brief proposal.

Your instructor will tell you which of the tasks you are to complete.

Get started on your job.

When You're Finished

Reflecting on This Case In a 250- to 500-word response to your instructor, discuss (a) what you learned from this case, (b) how you could relate this case to work situations you will face in your chosen career, and, if applicable, (c) the ways in which this case compares to similar situations you have already faced at work. Your instructor will tell you whether your response should be submitted as a memo, an e-mail, or a journal entry, or in a different format.

Moving beyond This Case Identify a health or safety issue affecting your community that interests you. Research this issue, and write a proposal suitable for submission to the Heathcot-Ann Foundation.

Task 1

●　●　●　●

Evaluate Proposals

Jessica hands you four proposals (**Documents 4.2** to **4.5**) to evaluate. "Before you start," Jessica says, "review the foundation's goals and evaluation criteria as well as a winning proposal from last quarter" (**Figure 4.1**, p. 38). She asks you to write the foundation's board a memo in which you rank the submissions from best to worst, list monetary awards, and provide a concise justification for your recommendations.

DOCUMENT 4.2

Skyview Coalition
Proposal

Contact Person: Sherilyn Baxter
Skyview Coalition
17 East Main Street
Boise, Idaho 83705

Problem Statement

Problem #1: We lack the funds to update our "Services Directory."

Problem #2: We lack the funds to serve a specific health need in our community: money for prescription medications.

Goals

Goal #1: Update and distribute at least 100 copies of a new "Services Directory" by January 31, 2013.
Goal #2: Meet specific funding needs for prescription medications that would otherwise be unmet.

Program Description

The Skyview Coalition (SC) is a partnership of King County community service groups. The SC serves as the central meeting place for groups and agencies to gather and coordinate the health-related needs of persons in King County. It has been the traditional role of the SC to coordinate services to prevent any at-risk person from falling through the cracks. The problem with this approach has always been the lack of an easy way to set out in clear terms which agency provides what services. SC members worked together to prepare a printed "Services Directory." This directory allowed people to locate services by category of need (for example, shelter, health care, etc.) This directory needs updating.

We also help persons who do not qualify for any type of assistance with the cost of prescription medications. In past years, we have been funded on a limited basis by the Renzo Foundation. However, our funding has evaporated when community need has sharply risen.

Budget Justification

Goal #1: Cost of supplies for new services directory:

Letter sized copy paper	10 reams @ $9.00 plus tax, shipping	$107.20
Copy machine toner cartridge	1 plus tax, shipping	$120.17
Three-ring binders	100 @ $3.50, plus tax, shipping	$378.67
Labels for binder covers	5 packages @ 3.99 plus tax, shipping	$25.45
Section dividers	100 packages @ 2.99 plus tax, shipping	$332.99
	Total:	**$965**

Goal #2: Prescription medication funding:

Prescription funding	3 months	$1000.00
	Total:	**$1000.00**

Note: This grant would by used for three months of prescription funding, and would allow us time to regroup.

Overall Total Grant Request: **$1,965.00**

DOCUMENT 4.3

Gameday Brigade
Proposal

Contact Person: Marc Heidelburg
Rehabilitation Unit
State Children's Hospital
3402 Valley Road
Garden Valley ID, 83704

The Problem: The Rehabilitation Unit (a.k.a. Rehab) at State Children's Hospital (SCH) is a sixteen bed extended care unit. Pediatric physical rehabilitation hospital wards train children who have been severely and permanently disabled to learn to live as active a life as each of them can and to varying extents, reincorporated them back into the mainstream of life. Most of these kids have closed head type injuries and/or spinal cord damage secondary to serious trauma and therefore need constant medical supervision that can only be achieved with long term hospitalization.

The Solution: *Gameday Brigade* proposes to orchestrate a "Take me out to the ballpark day" for the kids in Rehab, and those parents who would like to share in the event. The event is a field trip for the kids in Rehab at SCH to our town's minor league home baseball game.

My Approach: Marc Heidelburg (me) is an individual is not an employee of or permanently associated with State Childrens' Hospital. I have a BS in Biochemistry/Molecular Biology from the State University where I graduated Magna Cum Laude. I am in my final semester of the Master's in Chemistry program at SU. Furthermore, I am a former Pediatric Rehabilitation Volunteer at Skyview Teaching Hospital and a former medical student form the State College of Medicine.

How the Heathcot-Lynn Foundation Can Help: With the demonstrated commitment you organization has shown to making heath care possible for all community members, I am requesting approximately $500 to cover admission for the children, the support staff, and interested parents. If you approve this proposal request, I will use it as the foundation upon which to seek support from local businesses. Your consideration of this proposal project is greatly appreciated. On behalf of the Rehab children, thank you.

Contact Person: Robin Collins
Community Outreach, Recreation, and Education (CORE) Center
125 Weeping Willow Lane
Kuna, ID 83704

Need

In 2011, there were 136 live births to women, ages 15-19 in King County. The total population of women ages 15-19 in King County was 1,458 that year. One out of ten teenagers became parents in King County in 2010. Many teen parents have not had the opportunity to develop parenting skills which are essential to the health, safety, and education of their children. The Community Outreach, Recreation, and Education (CORE) Center has developed a Teen Initiative Plan (TIP) to deal with the problems of pregnant and parenting teens in King County.

Goals

Through the CORE Center, the King County Health Department and the Sweetwater School System have collaborated to provide prenatal care to pregnant teens at local schools. The State Department of Human Services has coordinated with the Health Department and the Sweetwater School System to develop an on-campus day care program for teen parents who are still in school. In addition to the day care, the CORE Center will provide TIP classes to teen parents whose children are enrolled in the on-campus day care center.

Approach

The TIP parenting classes will be required of teen parents whose children are enrolled in the on-campus day care program. The classroom program will be designed to:

- Teach the importance of good nutritional habits and preventative health care
- Teach the importance of child safety
- Teach responsible decision making
- Teach the importance of education

A CORE Center member will coordinate speakers for the various topics and lead class discussion. Ideally, the class would be conducted by a certified teacher who would invite guest speakers through the CORE Center. The TIP class would be an elective course for the student.

Budget Justification

1 – 32" LCD HDTV DVD Combo	$408
1 – Cart for Flat Panel TV	100
Parenting Textbooks	450
50 pupils textbooks	
3 teacher guides	
10 Classroom DVDs	550
Teaching aids	300
Total Grant	**1,808**

DOCUMENT 4.5

Defensive Driving Course
Proposal

Contact Person: Eric Glenn
State College Student Services
3476 Sam Peck Drive
Boise, ID 83704

Situation

According to the *Idaho Traffic Collision 2011 Report*, 26,241 traffic collisions totaling $1.6 billion occurred in 2011. Drivers between the ages of 20 and 34 represented a major percentage of drivers involved in those traffic collisions. Although they comprised 27% of all licensed drivers, they accounted for 34% of all collision-involved drivers and 35% of drivers involved in fatal or injury collisions. Indeed, traffic officers issued 154,649 traffic tickets in 2011. Half of the traffic violations were basic rule/speeding violations.

For most students, being involved in a traffic collision is depressing. The consequences of these matters vary from affecting grades to dropping out of school. Most collisions are avoidable; therefore I want to try to introduce a two-hour Defensive Driving Course (DDC) offered at eight different times for State College students.

Goal

The goal of the DDC is to improve students' driving skills.

Method

Teach four workshops each spring and fall semester. All students are welcome, but students who often have traffic collisions or violation tickets will have priority. The best way to illustrate the concepts of defensive driving is through hands-on driving exercises. To keep costs low, we will use a combination of discussion and video instruction. DDC is not new to Idaho. Many state employees attended an approved DDC during the past three years. After taking the DDC, state employee accident rates have declined, resulting in lower liability premiums for state agencies.

Budget

Each DDC will cost $220 ($1760 for all eight workshops). The cost covers instructor honorariums, course materials, audio-visual equipment fees, refreshments, and advertising.

Benefits

Conducting the DDC at State College will be a benefit to everyone in King County:

- Enhance participants' life and safety skills
- Reduce the human tragedy and property loss caused by motor vehicle accidents
- Increase the number of safe drivers in the county

Task 2

• • ● • •

Reflect on the Review Process

After you have evaluated the proposals (**Documents 4.2** to **4.5**, pp. 40–43), meet with three or four other reviewers. Discuss the proposals, and reach a consensus on each proposal: whether it should be funded and, if so, what monetary award should be given. Jessica is also interested in the effectiveness of the proposal-review process. After reaching a consensus, hold a short debriefing session in which you consider the following questions:

- How did you read each proposal? Did you read in a linear fashion from top to bottom, or did you skip around and read sections out of order? Did you skim or skip any parts?

- In what ways did each reviewer's background, personal beliefs, and interpretation of the evaluation criteria affect funding decisions?

- What aspects of the content, organization, and design of the proposals made reviewing them difficult? What made the task easier?

   Jessica would like to offer proposal writers some advice on the Heathcot-Ann Web site. Based on your discussion and experience as a reviewer, provide advice on how to write a successful proposal. "Don't worry about building a Web page. Just e-mail me a document with the content designed the way you want it to look online, and include notes to our Web designer regarding any additional online features you want included, such as links."

Task 3

• • ● • •

Respond to Proposals

The foundation's board of directors asks you to write an e-mail to each organization for whose proposal you recommended reduced or no funding (**Documents 4.2** to **4.5**). Jessica explains, "We don't want to discourage these groups. Many of our best proposals come from groups who were first rejected and who then resubmitted revised proposals." In your messages, remember to thank the individual for submitting a proposal, explain why the proposal received reduced or no funding, suggest ways of improving the proposal, and, if appropriate, encourage the organization to resubmit the proposal during the next quarter.

Task 4

• • • •

Write a Proposal

After serving as a social media intern at the foundation for a semester, you resign and accept a job as a technical writer for King County Public Safety Department. You learn that the department wants to start a program to supply safety helmets to children under the age of 16 who cannot afford them. You mention that the Heathcot-Ann Foundation might be a good funding source and volunteer to develop a proposal. Your supervisor, Rene Johnson, takes you up on your offer and sends you an e-mail with the information she has collected (**Document 4.6**). Using her notes and other sources, write a proposal suitable for submission. Also, in a reply message to Rene, briefly justify your rationale for including or omitting visuals.

DOCUMENT 4.6

E-mail with Notes on Bicycle Helmet Program

To: [your name]
Subject: Bicycle Helmet Proposal Notes

Thanks for agreeing to develop this proposal. With more than 1.7 million people every year in the United States suffering a traumatic brain injury (TBI), I think the best way we can help families address this serious public health issue is through education and prevention. To help you get started, I've included below some notes on the idea as well as some graphics you might choose to include in the proposal. I'm not sure if we should include visuals or if the ones I found are suitable. What do you think?

Thanks,
Rene

Although there is no federal law in the United States requiring bicycle helmets, King County adopted in 1997 a bicycle helmet law requiring youths 16 and under to wear an approved protective helmet. In fact, 14 states still have no state or local bicycle helmet laws.

In 2011, police officers in King County issued 205 tickets or warnings for helmet law violations. A common reason given by children for not wearing a helmet was that they did not own a helmet. Many also said they could not afford one.

We need to provide helmets to families who cannot afford them. If families cannot afford helmets, then they certainly cannot afford the medical bills if kids sustain head injuries. We could also give out "citations" to kids who are wearing helmets, encouraging positive behavior. I've attached an example from the Norwood Police Department.

In 2011, 19 children under the age of 16 were hospitalized from bicycle-related traumatic brain injuries in King County. In fact, according to the Maine Bureau of Highway Safety, "More children, ages 5 to 14, go to hospital emergency rooms for injuries associated with bicycles than with any other sport."

We should give each police officer brochures to give to minors who are not using helmets. These brochures should emphasize the risks they are taking and encourage them to wear helmets. We could also educate kids on the proper way to wear a helmet (see my second and third attachments). Correctly worn, a bicycle helmet can reduce the chance of head injury by up to 85 percent.

CONTINUED ➲

Police to issue coupons redeemable at fire or police stations for free helmets. Another option is to place a few helmets in each police vehicle so officers can hand them out right on the spot.

Approximately 8,500 children between the ages of 5 and 16 are living in King County today. Many already have helmets or can easily afford one, so we probably need funds for about 100 helmets. Need small, medium, and large helmets. Helmets need to be unisex in color and design.

If we can just get 100 children who otherwise would not wear a helmet to do so, we can make a difference. As more kids wear helmets, they'll see it's cool to wear a helmet.

Attachment 1: Norwood Police Citation

Attachment 2: Bicycle Helmet Fit

Attachment 3: Bicycle Helmet Fit

Reports
Learning to Write in an Organization

The Situation

You have accepted a technical-communication internship at the Washington, DC headquarters of the U.S. Energy Information Administration (EIA). The EIA, the statistical and analytical agency of the U.S. Department of Energy, provides data and information on energy sources. The EIA publishes reports, Web products, press releases, databases, and maps. Publications are issued weekly, monthly, quarterly, annually, and as needed or upon request.

As an intern, you will work closely with Merilee Summerfield, a technical publications editor. "Our Web site is the primary communication channel for the EIA," Merilee explains. "We've just launched the first phase of a comprehensive site redesign that will affect the way we write Web content. Using a uniform style throughout our site tells users that the EIA has high standards for our content and our data." She has asked you to develop Web content and to help scientists as they write various EIA documents.

When writing Web content, EIA writers follow recommendations from *The Energy Information Administration's Web Editorial Style Guide* (**Figure 5.1**). The guide specifies details of the writing style required by the EIA, including rules for punctuation, word usage, tone, and hypertext links. The guide also contains direction on matters related to EIA work specifically, such as editorial voice, capitalization of frequently used words and terms, abbreviation of units of measure, and notes.

The Challenge

Learning to write in a new organization involves a learning curve. You will experience part of this learning curve when you are asked to quickly learn about the EIA and what represents consistent, correct, readable Web content within that organization. You will also have to grapple with unfamiliar content. As you study model documents, consider their audience and purpose, become knowledgeable about their contents, and familiarize yourself with the EIA's *Web Editorial Style Guide*, you will learn how to adapt your writing to meet the specific needs of an organization.

Case 5: Reports

Figure 5.1
Cover of the EIA's *Web Editorial Style Guide*

Your Job

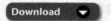

To download electronic copies of documents in this chapter, click on **Case 5 Documents** on bedfordstmartins.com /techdocs.

During your first week as an EIA intern, you will need to learn about EIA publications and Web editorial guidelines. As a new intern, you may be asked to do the following:

▶ Study a preface to a technical report to learn the characteristics of the organization's reports.

▶ Contrast a printed report to a report designed for online use, and evaluate their features.

▶ Learn about appropriate content and level of detail for an energy report.

▶ Use the organization's style guide to answer writing-related questions posed by analysts.

Your instructor will tell you which of the tasks you are to complete.

Get started on your job (see page 50).

When You're Finished

Reflecting on This Case   In a 250- to 500-word response to your instructor, discuss (a) what you learned from this case, (b) how you could relate this case to work situations you will face in your chosen career, and, if applicable, (c) the ways in which this case compares to similar situations you have already faced at work. Your instructor will tell you whether your response should be submitted as a memo, an e-mail, or a journal entry, or in a different format.

Moving beyond This Case   Locate a professional who spends a significant amount of his or her time writing at work and is willing to be interviewed. You may choose to interview someone from your chosen field or a different profession. Investigate how this professional learned about writing for his or her particular organization. You might ask the following:

▶ What types of documents do you regularly write for your employer?

▶ How did writing in college differ from writing in your organization?

▶ What did you need to learn about writing in your organization? How did you go about learning it?

▶ Did anyone within the organization help you? If so, who, and in what ways did they help?

▶ What advice do you have for new employees learning to write?

Present your findings in a report to your instructor.

Task 1

● ● ● ●

Learn about an Organization's Reports

First, Merilee would like you to learn about EIA, and, in particular, what EIA writers want to accomplish with their energy publications. She suggests that you start by reading "Our Work" under the heading "About EIA" on the EIA site at www.eia.gov/about/eia_explained.cfm (**Figure 5.2**) and explore other links on the site such as "Mission and Overview" (**Figures 5.3** and **5.4**). To introduce you

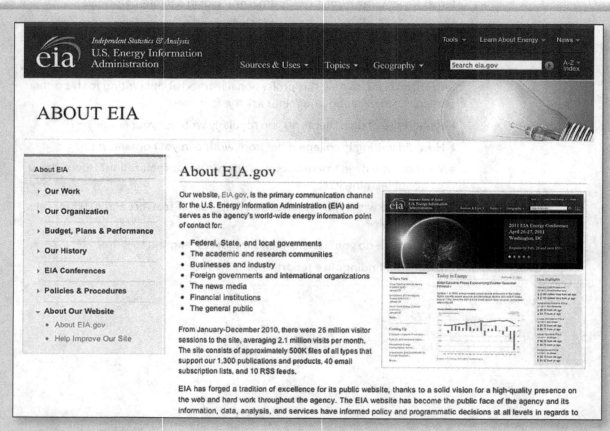

FIGURE 5.2
"About EIA" Page of the
EIA Web Site

to EIA publications, Merilee asks you to study a preface (**Document 5.5**) and write a two-page memo in which you do the following: (a) describe the preface's audience and purpose; (b) evaluate how well the preface explains the subject, purpose, background, and scope of the report; (c) describe and evaluate the preface's organizational structure; and (d) discuss the preface's level of formality and voice (active or passive). To get you started, Merilee has provided brief marginal notes.

FIGURE 5.3
"Mission and Overview"
Page of the EIA Web Site

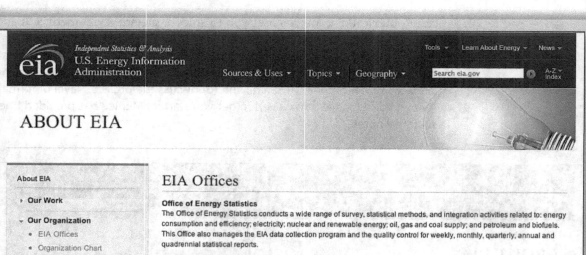

FIGURE 5.4
"EIA Offices" Page of the
EIA Web Site

DOCUMENT 5.5

Preface from *International Energy Outlook 2010*

> At the EIA, we label some of our introductory sections as prefaces and place them in the front matter of the report. Note lowercase Roman numeral pagination (e.g., ix and x). Other organizations place introductions in the body of the report.

Preface

> We provide our readers with a concise, one-sentence description of our report.

This report presents international energy projections through 2035, prepared by the U.S. Energy Information Administration, including outlooks for major energy fuels and associated carbon dioxide emissions.

> An overview of how data are grouped in the report.

The *International Energy Outlook 2010 (IEO2010)* presents an assessment by the U.S. Energy Information Administration (EIA) of the outlook for international energy markets through 2035. U.S. projections appearing in *IEO2010* are consistent with those published in EIA's *Annual Energy Outlook 2010 (AEO2010)* in April 2010.

The *IEO2010* projections are based to the extent possible on U.S. and foreign laws, regulations, and standards in effect at the start of 2010. The potential impacts of pending or proposed legislation, regulations, and standards are not reflected in the projections, nor are the impacts of legislation for which the implementing mechanisms have not yet been announced. In addition, mechanisms whose implementation cannot be modeled given current capabilities or whose impacts on the energy sector are unclear are not included in *IEO2010*. For example, the European Union's Emissions Trading System, which includes non-carbon dioxide emissions and non-energy-related emissions, are not included in this analysis.

> A statement of methods used to make projections.

IEO2010 focuses exclusively on marketed energy. Non-marketed energy sources, which continue to play an important role in some developing countries, are not included in the estimates.

> A statement of types of energy not included in this report.

The *IEO2010* consumption projections are grouped according to Organization for Economic Cooperation and Development membership. (OECD includes all members of the organization as of March 1, 2010, throughout all time series included in this report. Chile became a member on May 7, 2010, but its membership is not reflected in *IEO2010*.) There are three basic groupings of OECD countries: North America (United States, Canada, and Mexico); OECD Europe; and OECD Asia (Japan, South Korea, and Australia/New Zealand). Non-OECD is divided into five separate regional subgroups: non-OECD Europe and Eurasia, non-OECD Asia, Africa, Middle East, and Central and South America. Russia is represented in non-OECD Europe and Eurasia; China and India are represented in non-OECD Asia; and Brazil is represented in Central and South America. In some instances, the *IEO2010* production models have different regional aggregations to reflect the important producer regions (for example, Middle East OPEC is a key region in the projections of liquid supplies). The complete regional definitions are listed in Appendix M.

The report begins with a review of world trends in energy demand and the major macroeconomic assumptions used in deriving the *IEO2010* projections, which—

Objectives of the *IEO2010* Projections

The projections in *IEO2010* are not statements of what will happen, but what might happen given the specific assumptions and methodologies used for any particular scenario. The Reference case projection is a business-as-usual trend estimate, given known technology and technological and demographic trends. EIA explores the impacts of alternative assumptions in other scenarios with different macroeconomic growth rates and world oil prices. The *IEO2010* cases generally assume that current laws and regulations are maintained throughout the projections. Thus, the projections provide policy-neutral baselines that can used to analyze international energy markets.

While energy markets are complex, energy models are simplified representations of energy production and consumption, regulations, and producer and consumer behavior. Projections are highly dependent on the data, methodologies, model structures, and assumptions used in their development. Behavioral characteristics are indicative of real-world tendencies, rather than representations of specific outcomes.

Energy market projections are subject to much uncertainty. Many of the events that shape energy markets cannot be fully anticipated. In addition, future developments in technologies, demographics, and resources cannot be foreseen with certainty. Key uncertainties in the *IEO2010* projections for economic growth and oil prices are addressed through alternative cases.

EIA has endeavored to make these projections as impartial, reliable, and relevant as possible. They should, however, serve as an adjunct to, not a substitute for, a complete and focused analysis of public policy initiatives.

> Because this "disclaimer" is so important, we placed it in a shaded box.

CONTINUED ➡

DOCUMENT 5.5

(continued)

Beginning at the bottom of page ix, we include a detailed advance organizer in these three paragraphs.

for the first time—extend to 2035. In addition to Reference case projections, High Economic Growth and Low Economic Growth cases were developed to consider the effects of higher and lower growth paths for economic activity than are assumed in the Reference case. *IEO2010* also includes a High Oil Price case and, alternatively, a Low Oil Price case. The resulting projections—and the uncertainty associated with international energy projections in general—are discussed in Chapter 1, "World Energy Demand and Economic Outlook."

Projections for energy consumption and production by fuel—liquids (primarily petroleum), natural gas, and coal—are presented in Chapters 2, 3, and 4, along with reviews of the current status of each fuel on a worldwide basis. Chapter 5 discusses the projections for world electricity markets—including nuclear power, hydropower, and other commercial renewable energy resources—and presents forecasts of world installed generating capacity. Chapter 6 provides a discussion of industrial sector energy use. Chapter 7 includes a detailed look at the world's transportation energy use. Finally, Chapter 8 discusses the outlook for global energy-related carbon dioxide emissions.

Appendix A contains summary tables for the *IEO2010* Reference case projections of world energy consumption, gross domestic product, energy consumption by fuel, carbon dioxide emissions, and regional population growth. Summary tables of projections for the High and Low Economic Growth cases are provided in Appendixes B and C, respectively, and projections for the High and Low Oil Price cases are provided in Appendixes D and E, respectively. Reference case projections of delivered energy consumption by end-use sector and region are presented in Appendix F. Appendix G contains

summary tables of projections for world liquids production in all cases. Appendix H contains summary tables of Reference case projections for installed electric power capacity by fuel and regional electricity generation. Appendix I contains summary tables for projections of world natural gas production in all cases. Appendix J includes a set of tables for each of the four Kaya Identity components. In Appendix K, a set of comparisons of projections from the International Energy Agency's *World Energy Outlook 2009* with the *IEO2010* projections is presented. Comparisons of the *IEO2010* and *IEO2009* projections are also presented in Appendix K. Appendix L describes the models used to generate the *IEO2010* projections, and Appendix M defines the regional designations included in the report.

The *IEO2010* projections of world energy consumption were generated from EIA's World Energy Projections Plus (WEPS+) modeling system. WEPS+ is used to build the Reference case energy projections, as well as alternative energy projections based on different assumptions for GDP growth and fossil fuel prices. The *IEO2010* projections of global natural gas production and trade were generated from EIA's International Natural Gas Model (INGM), which estimates natural gas production, demand, and international trade by combining estimates of natural gas reserves, natural gas resources and resource extraction costs, energy demand, and transportation costs and capacity in order to estimate future production. The Generate World Oil Balance (GWOB) application is used to create a "bottom up" projection of world liquids supply—based on current production capacity, planned future additions to capacity, resource data, geopolitical factors, and oil prices—and to generate conventional crude oil production cases.

This last paragraph includes more discussion methods. Do you think ending the preface with a paragraph on methods is effective?

Task 2

Compare and Evaluate Report Designs

To help you better understand how the communication medium will affect the way the content is presented, Merilee asks you to take a closer look at three different designs for the organization's annual *International Energy Outlook* (*IEO*) report (**Documents 5.6** to **5.8**). In the margins, she has posed questions to help you focus on the different designs. Respond to her questions in a brief memo.

DOCUMENT 5.6

Web Page of Most Recent *International Energy Outlook* Report

What navigation aids are available on this page to help readers quickly find the content they seek?

What types of readers might use the hyperlinks to data tables and the Interactive Table Viewer?

In what ways are these links to related content useful to readers?

DOCUMENT 5.7

Web Page of Archived
*International Energy
Outlook 2010*

This page displays an archived *IEO* report. In what ways does this page display content and related resources differently from the page displaying the most recent report (Document 5.6)?

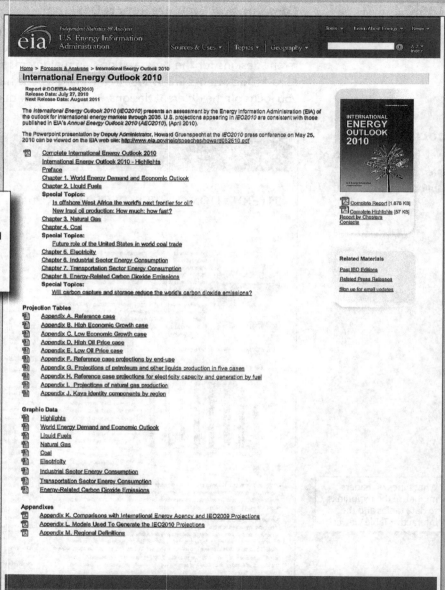

DOCUMENT 5.8

Highlights Page from Printed Version of Archived *International Energy Outlook 2010*

What navigation aids are not available in a printed report?

Highlights

World marketed energy consumption increases by 49 percent from 2007 to 2035 in the Reference case. Total energy demand in non-OECD countries increases by 84 percent, compared with an increase of 14 percent in OECD countries.

In the *IEO2010* Reference case, which does not include prospective legislation or policies, world marketed energy consumption grows by 49 percent from 2007 to 2035. Total world energy use rises from 495 quadrillion British thermal units (Btu) in 2007 to 590 quadrillion Btu in 2020 and 739 quadrillion Btu in 2035 (Figure 1).

The global economic recession that began in 2008 and continued into 2009 has had a profound impact on world energy demand in the near term. Total world marketed energy consumption contracted by 1.2 percent in 2008 and by an estimated 2.2 percent in 2009, as manufacturing and consumer demand for goods and services declined. Although the recession appears to have ended, the pace of recovery has been uneven so far, with China and India leading and Japan and the European Union member countries lagging. In the Reference case, as the economic situation improves, most nations return to the economic growth paths that were anticipated before the recession began.

How has the writer designed this page to be printed and read?

The most rapid growth in energy demand from 2007 to 2035 occurs in nations outside the Organization for Economic Cooperation and Development[1] (non-OECD nations). Total non-OECD energy consumption

increases by 84 percent in the Reference case, compared with a 14-percent increase in energy use among OECD countries. Strong long-term growth in gross domestic product (GDP) in the emerging economies of non-OECD countries drives their growing energy demand. In all non-OECD regions combined, economic activity—as measured by GDP in purchasing power parity terms—increases by 4.4 percent per year on average, compared with an average of 2.0 percent per year for OECD countries.

The *IEO2010* Reference case projects increased world consumption of marketed energy from all fuel sources over the 2007-2035 projection period (Figure 2). Fossil fuels are expected to continue supplying much of the energy used worldwide. Although liquid fuels remain the largest source of energy, the liquids share of world marketed energy consumption falls from 35 percent in 2007 to 30 percent in 2035, as projected high world oil prices lead many energy users to switch away from liquid fuels when feasible. In the Reference case, the use of liquids grows modestly or declines in all end-use sectors except transportation, where in the absence of significant technological advances liquids continue to provide much of the energy consumed.

What interactive features are not available in a printed report?

Figure 1. World marketed energy consumption, 2007-2035 (quadrillion Btu)

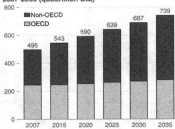

Figure 2. World marketed energy use by fuel type, 1990-2035 (quadrillion Btu)

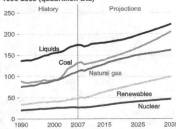

[1]Current OECD member countries (as of March 10, 2010) are the United States, Canada, Mexico, Austria, Belgium, Czech Republic, Denmark, Finland, France, Germany, Greece, Hungary, Iceland, Ireland, Italy, Luxembourg, the Netherlands, Norway, Poland, Portugal, Slovakia, Spain, Sweden, Switzerland, Turkey, the United Kingdom, Japan, South Korea, Australia, and New Zealand. Chile became a member on May 7, 2010, but its membership is not reflected in *IEO2010*.

U.S. Energy Information Administration / International Energy Outlook 2010   1

Task 3

• • ● •

Learn about Appropriate Report Content

To help you gain a better understanding of appropriate content and level of detail when reporting energy information, Merilee asks you to take a closer look at three excerpts focusing on carbon dioxide emissions (**Documents 5.9** to **5.11**) from a report on emissions of greenhouse gases in the United States. In the margins, she has posed questions to help you focus on the characteristics of the selections. Respond to her questions in a brief memo.

DOCUMENT 5.9

Page from *Greenhouse Emissions* Report

> Study the text sections on the next three pages. What types of information are discussed in these sections?

> Why is it important for readers to understand the key factors contributing to the decrease in carbon dioxide emissions?

> Why do you think our reports rely so heavily on graphics?

---

2. Carbon dioxide emissions

2.1. Total carbon dioxide emissions

Annual U.S. carbon dioxide emissions fell by 419 million metric tons in 2009 (7.1 percent), to 5,447 million metric tons (Figure 9 and Table 6). The annual decrease—the largest over the 19-year period beginning with the 1990 baseline—puts 2009 emissions 608 million metric tons below the 2005 level, which is the Obama Administration's benchmark year for its goal of reducing U.S. emissions by 17 percent by 2020.

The key factors contributing to the decrease in carbon dioxide emissions in 2009 included an economy in recession with a decrease in gross domestic product of 2.6 percent, a decrease in the energy intensity of the economy of 2.2 percent, and a decrease in the carbon intensity of energy supply of 2.4 percent.

Energy-related carbon dioxide emissions accounted for 98 percent of U.S. carbon dioxide emissions in 2009 (Table 6) when adjusted for bunker fuels and U.S. Territories. The predominant share of carbon dioxide emissions comes from fossil fuel combustion, with smaller amounts from the nonfuel use of energy and emissions from U.S. Territories and international bunker fuels. Other relatively small sources include emissions from industrial processes, such as cement and limestone production.

U.S. carbon dioxide emissions, 1990, 2005, 2008, and 2009

	1990	2005	2008	2009
Estimated emissions (million metric tons)	5,040.9	6,055.2	5,865.5	5,446.8
Change from 1990 (million metric tons)		1,014.3	824.6	405.9
(percent)		20.1%	16.4%	8.1%
Average annual change from 1990 (percent)		1.2%	0.8%	0.4%
Change from 2005 (million metric tons)			-189.7	-608.4
(percent)			-3.1%	-10.0%
Change from 2008 (million metric tons)				-418.7
(percent)				-7.1%

Figure 9. Annual change in U.S. carbon dioxide emissions, 1991-2009

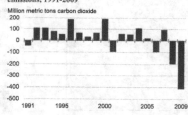

Million metric tons carbon dioxide

Table 6. U.S. carbon dioxide emissions from energy and industry, 1990-2009 (million metric tons carbon dioxide)

Fuel type or process	1990	1995	2000	2003	2004	2005	2006	2007	2008	2009
Energy consumption										
Petroleum	2,186.6	2,207.1	2,460.6	2,518.4	2,608.6	2,627.6	2,602.5	2,603.2	2,443.5	2,318.8
Coal	1,821.4	1,913.1	2,155.5	2,135.7	2,160.2	2,181.9	2,146.9	2,172.2	2,139.4	1,876.8
Natural gas	1,024.6	1,183.7	1,240.6	1,191.1	1,194.4	1,175.2	1,157.0	1,234.7	1,243.0	1,218.0
Renewables[a]	6.2	10.3	10.5	11.8	11.5	11.6	11.9	11.7	12.0	12.0
Energy subtotal	*5,038.7*	*5,314.3*	*5,867.2*	*5,856.9*	*5,974.7*	*5,996.4*	*5,918.3*	*6,021.8*	*5,838.0*	*5,425.6*
Nonfuel use emissions[b]	94.1	101.9	105.5	97.3	105.0	100.7	103.5	101.7	97.7	82.8
Nonfuel use sequestration[c]	250.0	283.6	306.0	290.0	313.4	303.5	299.8	292.1	264.4	245.7
Adjustments to energy[d]	-82.9	-63.2	-64.7	-32.6	-45.3	-44.6	-62.7	-67.5	-76.1	-66.0
Adjusted energy subtotal	*4,955.9*	*5,251.1*	*5,802.6*	*5,824.3*	*5,929.3*	*5,951.8*	*5,855.7*	*5,954.2*	*5,761.9*	*5,359.6*
Other sources	85.1	102.3	97.8	98.9	102.0	103.5	105.9	105.3	103.6	87.3
Total	5,040.9	5,353.4	5,900.3	5,923.3	6,031.3	6,055.2	5,961.6	6,059.5	5,865.5	5,446.8

[a]Includes emissions from electricity generation using nonbiogenic municipal solid waste and geothermal energy.
[b]Emissions from nonfuel uses are included in the energy subtotal above.
[c]The energy content of nonfuel uses in which carbon is sequestered is subtracted from energy consumption before emissions are calculated.
[d]Adjustments include adding emissions from U.S. Territories and subtracting emissions from international bunker fuels, in keeping with international practices.
Note: Totals may not equal sum of components due to independent rounding.

U.S. Energy Information Administration | Emissions of Greenhouse Gases in the United States 2009   21

DOCUMENT 5.10

Excerpt on Energy-Related Carbon Dioxide Emissions from *Greenhouse Emissions* Report

Identify the strategies the writers use to integrate the graphics with the text.

What types of information do the graphics on this page communicate more effectively than the text above them?

Because our writers frequently use tables, I'd like you to become familiar with the parts of a table. Can you identify and list elements that make these tables effective?

Carbon dioxide

2.2. Energy-related carbon dioxide emissions

Energy-related carbon dioxide emissions account for more than 80 percent of U.S. greenhouse gas emissions. These emissions were down by 7.1 percent from 5,838 million metric tons in 2008 to 5,426 million metric tons in 2009. EIA breaks energy use into four end-use sectors (Table 7), and emissions from the electric power sector are attributed to the end-use sectors based on electricity sales to each sector. Growth in energy-related carbon dioxide emissions since 1990 has resulted largely from increases associated with electric power generation and transportation fuel use. All other energy-related carbon dioxide emissions (from direct fuel use in the residential, commercial, and industrial sectors) have been either flat or declining in recent years (Figure 10). In 2009, however, emissions from both electric power and transportation fuel use were down—by 9.0 percent and 4.3 percent, respectively—continuing a trend from 2008.

Reasons for the long-term growth in electric power and transportation sector emissions include: population growth; increased demand for electricity for computers and electronics in homes and offices; strong growth in demand for commercial lighting and cooling; substitution of new electricity-intensive technologies, such as electric arc furnaces for steelmaking in the industrial sector; and increased travel as a result of relatively low fuel prices and robust economic growth in the 1990s and early 2000s. Likewise, the recent declines in emissions from both the transportation and electric power sectors are tied to the economy, with people driving less and consuming less electricity over the years 2008 and 2009.

Other sources of U.S. energy-related carbon dioxide emissions have remained constant or declined, for reasons that include increased efficiencies in heating technologies, declining activity in older "smokestack" industries (such as steel, paper, and chemicals), and the growth of less energy-intensive industries, such as computers and electronics.

U.S. energy-related carbon dioxide emissions, 1990, 2005, 2008, and 2009

	1990	2005	2008	2009
Estimated emissions (million metric tons)	5,038.7	5,996.4	5,838.0	5,425.6
Change from 1990 (million metric tons)		957.7	799.2	386.9
(percent)		*19.0%*	*15.9%*	*7.7%*
Average annual change from 1990 *(percent)*		*1.2%*	*0.8%*	*0.4%*
Change from 2005 (million metric tons)			-158.4	-570.8
(percent)			*-2.6%*	*-9.5%*
Change from 2008 (million metric tons)				-412.4
(percent)				*-7.1%*

Figure 10. Energy-related carbon dioxide emissions for selected sectors, 1990-2009

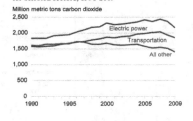

Million metric tons carbon dioxide

Electric power
Transportation
All other

Table 7. U.S. energy-related carbon dioxide emissions by end-use sector, 1990-2009 (million metric tons carbon dioxide)

Sector	1990	1995	2000	2003	2004	2005	2006	2007	2008	2009
Residential	963.4	1,039.1	1,185.1	1,230.1	1,227.8	1,261.5	1,192.0	1,242.0	1,229.0	1,162.2
Commercial	792.6	851.4	1,022.0	1,036.0	1,053.5	1,069.0	1,043.4	1,078.6	1,073.5	1,003.6
Industrial	1,695.1	1,742.8	1,788.1	1,691.9	1,731.1	1,675.2	1,661.1	1,661.6	1,597.6	1,405.4
Transportation	1,587.7	1,681.0	1,872.0	1,898.9	1,962.3	1,990.7	2,021.9	2,039.6	1,937.9	1,854.5
Total	5,038.7	5,314.3	5,867.2	5,856.9	5,974.7	5,996.4	5,918.3	6,021.8	5,838.0	5,425.6
Electricity generation[a]	1,831.0	1,960.1	2,310.2	2,319.2	2,351.5	2,416.9	2,359.5	2,425.9	2,373.7	2,160.3

[a]Electric power sector emissions are distributed across the end-use sectors. Emissions allocated to sectors are unadjusted for U.S. Territories and international bunker fuels. Adjustments are made to total emissions only.
Note: Totals may not equal sum of components due to independent rounding.

DOCUMENT 5.11

Excerpt on Residential-
Sector Carbon Dioxide
Emissions from
Greenhouse Emissions
Report

How would you describe
the level of detail and tone
of the text? Why do you
think the writers included
the details that they did?

Why do you think the
writers used two bulleted
lists here?

Based on these three pages,
describe how graphics
and text are arranged on
a page.

Carbon dioxide

2.3. Residential sector carbon dioxide emissions

Residential sector carbon dioxide emissions originate primarily from:

- Direct fuel consumption (principally, natural gas) for heating and cooking
- Electricity for cooling (and heating), appliances, lighting, and increasingly for televisions, computers, and other household electronic devices (Table 8).

Energy consumed for heating and cooling in homes and businesses has a large influence on annual fluctuations in energy-related carbon dioxide emissions because of variability in the weather as measured by heating and cooling degree-days. In 2009, heating degree-days were down slightly from 2008 (Figure 11). Although annual changes in cooling degree-days have a smaller impact on energy demand, the 4-percent decrease in 2009 helped to reduce emissions further.

In the longer run, residential emissions are affected by population growth, income, and other factors. From 1990 to 2009:

- Residential sector carbon dioxide emissions grew by an average of 1.0 percent per year.
- U.S. population is estimated to have grown by an average of about 1.1 percent per year.
- Income per capita (measured in constant dollars) grew by an average of 1.4 percent per year.
- Energy efficiency improvements for homes and appliances offset much of the growth in the number and size of housing units. As a result, direct emissions of carbon dioxide from the consumption of petroleum, coal, and natural gas in the residential sector in 2009 were up by only 0.9 percent from the 1990 level.

Residential sector carbon dioxide emissions, 1990, 2005, 2008, and 2009

	1990	2005	2008	2009
Estimated emissions (million metric tons)	963.4	1,261.5	1,229.0	1,162.2
Change from 1990 (million metric tons)		298.1	265.6	198.8
(percent)		30.9%	27.6%	20.6%
Average annual change from 1990 *(percent)*		1.8%	1.4%	1.0%
Change from 2005 (million metric tons)			-32.5	-99.3
(percent)			-2.6%	-7.9%
Change from 2008 (million metric tons)				-66.8
(percent)				-5.4%

Figure 11. Annual changes in U.S. heating degree-days and residential sector carbon dioxide emissions from direct fuel combustion, 1990-2009

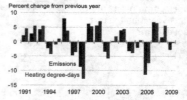

Percent change from previous year

Table 8. U.S. carbon dioxide emissions from residential sector energy consumption, 1990-2009 (million metric tons carbon dioxide)

Fuel	1990	1995	2000	2003	2004	2005	2006	2007	2008	2009
Petroleum										
Liquefied petroleum gas	22.2	24.9	35.0	34.3	32.3	32.3	28.1	30.5	34.9	36.5
Distillate fuel	71.6	66.2	66.2	66.2	67.6	62.5	52.1	53.1	48.5	44.5
Kerosene	4.6	5.4	6.8	5.1	6.1	6.1	4.8	3.2	1.5	1.9
Petroleum subtotal	98.4	96.5	108.0	105.6	106.0	100.9	85.0	86.8	84.9	82.9
Coal	3.0	1.7	1.1	1.2	1.1	0.8	0.6	0.7	0.7	0.6
Natural gas	238.3	262.9	270.8	276.4	264.3	262.4	237.5	257.3	265.8	259.1
Electricity[a]	623.7	678.1	805.2	846.9	856.4	897.3	868.9	897.2	877.5	819.5
Total	**963.4**	**1,039.1**	**1,185.1**	**1,230.1**	**1,227.8**	**1,261.5**	**1,192.0**	**1,242.0**	**1,229.0**	**1,162.2**

[a]Share of total electric power sector carbon dioxide emissions weighted by sales to the residential sector.
Note: Totals may not equal sum of components due to independent rounding.

U.S. Energy Information Administration | Emissions of Greenhouse Gases in the United States 2009   23

Task 4

• • • •

Use an Organization's Style Guide

Merilee would like your help responding to two e-mails (**Documents 5.12** and **5.13**) from EIA analysts. Because the analysts follow the EIA's *Web Editorial Style Guide* when writing reports, she wants you to consult relevant sections of the guide when responding to their queries. In both forwarded messages, she directs you to the appropriate sections of the guide. The *Web Editorial Style Guide* is available online at http://205.254.135.24/about/EIAWebEditorialStyleGuide.pdf. Write Merilee an e-mail with your responses to each message.

DOCUMENT 5.12

E-mail Query on How to Develop a Preface

To: [your name]
Subject: Doug Smith's Preface Question

Because we discussed an EIA preface earlier this week, I thought you could help me. How do you think we should respond to Doug? I recommend you use the *IEO 2010* preface as a model as well as information in the *Web Editorial Style Guide*, especially the Introduction and Chapters 1 and 12. E-mail me your response.

<<I've been asked to help write an upcoming report titled *Trends in Renewable Energy Consumption and Electricity 2012*. Because I haven't written one of these reports before, I have been reading past reports. However, I'm having a hard time figuring out how to write the preface to the report. I've pasted below the preface to the 2009 report. I'd like to write a longer, more detailed preface and format it for our Web site. Would you suggest how to do so?

Thanks.
Doug

Preface
The U.S. Energy Information Administration (EIA) reports detailed historical data on renewable energy consumption and electricity annually in its report, the Renewable Energy Annual. This report, *Trends in Renewable Energy Consumption and Electricity 2009*, provides an overview and tables with historical data spanning as far back as 1989 through 2009, including revisions. These tables correspond to identical tables to be presented in Chapter 1 of the Renewable Energy Annual 2009 and are numbered accordingly. The renewable energy resources in the report include: biomass (wood and derived fuels, municipal solid waste (MSW) biogenic, landfill gas, ethanol, biodiesel and other biomass); geothermal; wind; solar (solar thermal and photovoltaic); and conventional hydropower. Hydroelectric pumped storage is excluded, because it is usually based on non-renewable energy sources. Prior editions of this report may be found on the EIA Web site at www.eia.gov/renewable/annual. Definitions for terms used in this report can be found in the EIA's Energy Glossary at www.eia.gov/tools/glossary/index.cfm. General information about all the EIA surveys with data related to renewable energy and referenced in this report can be found here: www.eia.gov/survey.>>

DOCUMENT 5.13

E-mail Query on How to
Format a Table

To: [your name]
Subject: Marilyn Johnson's Table Questions

Marilyn needs help formatting a fairly simple table. Before you assist her, refer to the *Web Editorial Style Guide*, which discusses footnotes and notes in Chapter 11 and hypertext links in Chapter 12. You also might want to read Chapter 8 on abbreviations. E-mail me your responses to Marilyn's questions as well as a revised table reflecting your suggested revisions.

<<The attached table is part of the report *Short-Term Energy Outlook* I'm writing up for publication on the Web. I'm relatively new around here, and I'm still learning the EIA style. I need some help formatting my footnotes, notes, and hypertext links. Don't worry about understanding the science—I need your help getting it to look right. I have some specific questions:

1. I want to include the following footnotes: (1) West Texas Intermediate, (2) Average regular pump price, (3) On-highway retail, and (4) U.S. Residential average. Where should they go? I also want to include hyperlinks to our EIA online glossary (www.eia.gov/tools/glossary/index.cfm) for terms used in footnotes 1 and 3.

2. I'd like to include the source of my data (*Short-Term Energy Outlook*, September 2011) and include a hyperlink to the full report. How should I go about doing this?

3. I've used superscript 1 and italics to indicate projected prices in 2012 and 2013. What do you think?

Cheers,
Marilyn>>

Energy Source	Year			
	2010	**2011**	**2012[1]**	**2013[1]**
WTI Crude (1) ($/barrel)	61.65	79.40	*94.40*	*94.50*
Gasoline (2) ($/gallon)	2.35	2.78	*3.56*	*3.54*
Diesel (3) ($/gallon)	2.46	2.99	*3.85*	*3.87*
Heating Oil (4) ($/gallon)	2.52	2.97	*3.74*	*3.95*
Natural Gas (4)	12.12	11.19	*11.23*	*11.87*
Electricity (4) (cents/kilowatt hour)	11.51	11.58	*11.84*	*11.92*

[1] Prices are projections based on current data

Table 1. Price Summary of Major Energy Sources for 2010–2011 and Projections for 2012–2013.

Case 6

Instructions
Guiding Readers in Performing a Task

The Situation

Turnertronics Biomedical Systems (TBS) offers a line of anesthesia machines and monitoring devices for use in surgery and critical care. Anesthesia machines deliver medical gases to patients at a preselected mixture, pressure, and flow rate. Widely used in a variety of clinical settings, TBS's monitoring devices capture and display patients' vital measurements such as heart rhythm, blood pressure, and level of oxygen in blood (**Figure 6.1**).

The mission of TBS is to provide the most cost-effective and innovative biomedical equipment to the medical community. Key to achieving this mission is the company's Priti all-in-one monitoring systems. Used by anesthesia providers to both interpret patients' conditions and implement various treatments during care, the Priti monitors offer rapid measurements of all vital signs along with an easy-to-use touch screen.

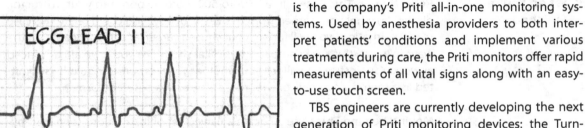

Figure 6.1
Engineer's Sketch of ECG
Lead II Waveform Display

TBS engineers are currently developing the next generation of Priti monitoring devices: the Turnertronics Priti5 All-in-One Anesthesia Monitoring System. Currently, the Priti5 exists as an early-stage prototype — so early, in fact, that the user interface design is still under development. For now, the engineers have provided you with a "best-guess" sketch of how the interface might look. Later, engineers will develop a touch-screen interface for the Priti5. As the company's documentation specialist, you are responsible for producing the documentation for the Priti5.

The Challenge

Often, documentation is developed at the same time as the product. In the early stages of product development, information is often scarce and product features are documented or revised as they are developed. Your challenge is to use available resources to document how the prototype works for now, knowing that you will have gaps in your understanding and that the prototype (and your initial documentation) will likely be different from the final product.

Your Job

As the documentation specialist on the project, you will need to analyze your audience and devise a way to show them how to use the new monitor correctly. Using the engineers' notes, you will start writing the step-by-step instructions for the Priti5 prototype. After being introduced to the project, you may be asked to do the following:

▶ Plan an interview with a subject-matter expert (SME).

▶ Analyze an engineer's notes, and evaluate a set of instructions.

▶ Use an engineer's notes to write step-by-step instructions.

▶ Develop an online training module for the monitor.

Your instructor will tell you which of the tasks you are to complete.

Get started on your job.

When You're Finished

Reflecting on This Case   In a 250- to 500-word response to your instructor, discuss (a) what you learned from this case, (b) how you could relate this case to work situations you will face in your chosen career, and, if applicable, (c) the ways in which this case compares to similar situations you have already faced at work. Your instructor will tell you whether your response should be submitted as a memo, an e-mail, or a journal entry, or in a different format.

Moving beyond This Case   Locate a flawed set of instructions for a step-by-step procedure, and revise them so that the directions are more effectively communicated. Or write instructions for a procedure that does not have any but needs them. Shareware programs and free software often feature imperfect instructions or no instructions at all. Shareware sites (such as www .shareware.com) might provide you with a good sample for this task.

Task 1

● ● ● ●

Plan an Interview with a Subject-Matter Expert

The Priti5 project manager, Miranda Gutierrez, wants you to learn about the potential users of the monitor and its features by interviewing various subject-matter experts — the engineers, sales representatives, and consulting anesthetists who best know the prototype and how it will be used. Miranda explains that these folks are busy and often do not recognize the importance of documentation. "You'll need to convince them to take time away from product development, sales, or patient care to talk with you. You will also need to ask appropriate questions that won't be seen as a waste of their time," she explains. As an example of what not to do, Miranda hands you an annotated copy of an e-mail (**Document 6.2**) that she received from a person in another division whom she was not inclined to help. Consequently, she asks you to send her a draft of the e-mail you will send to the SMEs to schedule an interview as well as the list of questions you plan to ask about audience, features, and medical terms. Working with two to three other students, brainstorm a list of questions about the Priti5 to ask each of the three different types of SMEs listed above, and then write the e-mails.

DOCUMENT 6.2

E-mail Requesting Information on the Priti4.3 Monitor

> The message projects the attitude that I exist solely to help Jeff.

To: Miranda Gutierrez
Subject: Priti4.3

I just realized I need information about the Priti4.3 before I leave. (I assume you worked on the 4 series.) Can the 4 recognize the use of the anesthetic agent Isoflurane? Can the monitor display two different ECG leads? Also, tell me something about the 4's other features.

Jeff

> The questions are flawed. By requiring only a "yes" or "no" response, Jeff won't learn very much. Also, the last question is too broad.

Task 2

• • • •

Evaluate Instructions

Miranda shows you the engineer's sketch of the touch-screen interface (**Figure 6.3**) and a set of basic instructions for using the Priti5 (**Document 6.4**). The final version of the Priti5 will have a touch-screen control panel rather than mechanical buttons and switches. The only exception is the on/off button. "I drafted instructions for when the machine is started. I think we should have these 'start-up' instructions printed on the side of the monitor as well as in the user guide. What do you think?" Write Miranda an e-mail evaluating the instructions and suggesting improvements.

FIGURE 6.3

Engineer's Sketch of Monitor User Interface

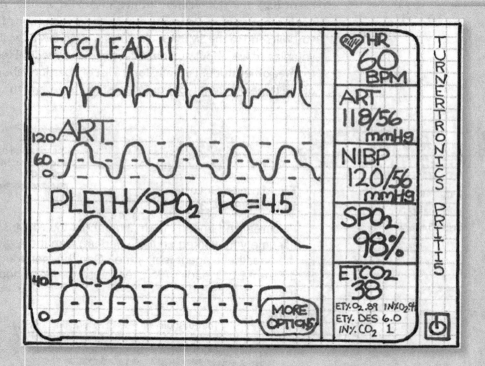

DOCUMENT 6.4

Basic Instructions for
Operating Priti5 Monitor

Introducing the PRITI5 All in One Anesthesia Monitoring System, manufactured by Turnertronics Biomedical Systems!

Basic Instructions for Using the PRITI5 All in One

Since the advent of the PRITI monitoring systems (prior versions 1-4), we have strived to include a wide range of options, allowing for optimal visualization and accurate data regarding any patient. In order for the anesthesia provider to interpret and implement various changes in the patient status, they must first be able to accurately analyze the data. By using the PRITI5 All in One, these tasks can be completed accurately and quickly due to the changes made in this new and upgraded monitor. Here's how:

1. Verify all connections to electrical, oxygen, air and vacuum systems with the Biomedical department. Turn monitor on. The power button is located in the front of the machine on the right lower corner.

2. Attach monitoring devices to patient.

3. If using simulation equipment, attach at this time.

4. Select the patient type by touching the "more options" button at bottom of the monitor screen until the scroll-down selection appears and touch adult, pediatric, or neonate to set the monitor.

5. Verify that appropriate waveforms—waveforms are essentially pictures on the screen that depict different measurements of heart rate and rhythm, blood pressure, and oxygenation at the capillary level, for starters—are present and accurate.

The upgrades to the monitor make it unique for use while monitoring an anesthetized patient having surgery, but can also be functional in any Critical Care Unit or Emergency Department.

Warning! Before using the monitor, the provider should review all attached materials and instructions included.

NOTE: Turnertronics is happy to provide one-on-one servicing of the PRITI5 All in One Anesthesia Monitoring System and recommends this be completed prior to use on patients.

Task 3

● ● ● ●

Write Instructions

A project engineer would like you to start documenting how to monitor several different vitals as well as how to analyze the anesthetic gas. She has given you her notes (**Document 6.5**), which Miranda has annotated for you. Using the engineer's notes and Miranda's comments, write the user guide instructions for the five features described in the notes.

DOCUMENT 6.5

Notes for Operating Priti5 Monitor

> Do operators need this information?

> Step for operators

> Feedback statement

> Do operators need this information?

> This procedure would be easier to understand if it was formatted as a numbered list.

> Where should we place this information?

ECG Monitoring

ECG monitoring = a waveform on the screen, shows how the electrical conduction through the heart creates a certain type of rhythm
- Can be reflected in different ways (depends on how the anesthetist/care provider places the electrode patches on the patient's chest)
- Detects aberrant or abnormal rhythms on the screen. May require advance knowledge of such rhythms in order to physically treat them.
- Both 3 and 5 lead ECG monitor cables and functions available on the PRITI5.

How to: To choose which leads to monitor, press or just touch the ECG waveform. Several options to change the reading on the screen will appear. Choose the option for *change lead* and then choose which leads to monitor in.

Background=Lead II and V are common with a 5 lead system (3 and 5 leads refer to the number of electrodes paced on the patient and the different views the monitor 'picks' up from the heart's electrical conduction), but choose best for patient & *visualization* (ask consultants if this term is used by care providers) of the rhythm.

Choose the option for *change size* and then either increase or decrease the size of the waveform. Choose the color option to visualize the waveform on the monitor. Several choices available. Select the option for ST segment monitoring, if desired and appropriate for the case and patient type.

Other selections print rhythm strips and display delayed recordings of changes to the rhythm.

IMPORTANT Alarm values/volumes come preset in the PRITI5.
To change default alarms and/or volumes: touch the screen display either on the individual waveform/readout or the right lower corner to bring up the pop-up list.

CONTINUED ➲

DOCUMENT 6.5

(continued)

How much of this is important to operators? Can we move some of this information to a different section in the user guide?

Press, select, touch: we need to consistently use the same word.

Should this information be a note?

Can we make this more concise?

Non Invasive Blood Pressure Monitoring (NIBP)

Non-invasive = external device. NIBP measures blood pressure with a cuff on the patient's arm or leg.

- Normal BP 120/80 for adult sized, (top number is systolic pressure & bottom number is diastolic)
- Pediatric patients = lower blood pressures & require different sized cuffs for accurate measurement.
- Anesthesia delivery can alter blood pressure significantly: emphasize the importance of a proper measurement

How to: Select appropriate BP cuff size for the patient & attach the cable to the cuff. Touch NIBP selection on the screen & choose: manual or automatic readings. Select the NIBP selection on the screen & set the time limits to activate the cuff.

Press start and reading will appear within 15 seconds on the screen. If using the cuff as a tourniquet, activate the venous stasis button under the NIBP selection using the touch screen to get there. Don't forget to deactivate the cuff when completed with the task

Arterial Line Monitoring

= form of monitoring BP:

- Anesthesia person inserts IV-like catheter into the radial artery (wrist?) & attaches it to a special tubing and cable system that produces a specific picture on the monitor screen = more accurate reading of the BP.

- Emphasize remembering to recalibrate the system to each patient when attaching line to the patient. Arterial line, Central Venous, and Pulmonary Artery waveforms: all can be interchanged on the screen depending on the view required for accurate monitoring of the specific patients.

How to:

Attach appropriate module & cable to the monitoring system. Arterial line waveform section will appear on monitor screen with placement of module. Select the *arterial line* (ART) by touch screen to set label for the screen. Options include ALINE, ABP and ART. Select the color choice & size of the waveform using touch screen. Calibrate or "zero" the waveform using hospital system procedures & opening the line to air while touching the "zero" button on the screen. The monitor verifies the calibration with an audio tone and message on screen below the waveform.

CONTINUED ➡

Oximeter Monitoring

= measures oxygen delivery at the capillary level & depicted by number on the screen from 0-100%.
- Normal readings = 95-100%, but can be lower if patient has pulmonary disease or requires O_2
- Measured on an extremity (finger, toe, ear). Affected by low perfusion & temperature of extremity.
- Oximeter one of the most important monitors in anesthesia because it gives advance warning regarding changes in the patient's oxygen levels.

How to:
- Attach oximeter device to patient. Several different types of attachments available, including finger, toe, nasal, ear, forehead and wrist depending on the needs of the patient. Also, different sizes are available for all patients.
- Visualize the waveform. Perfusion capability = a number that appears on the screen to tell the anesthesia provider how well the device will read the patient's oxygen values based on the perfusion of the extremity. PC through the oximeter will be noted on the screen. Identified through numerical reading. A PC of 1 = poor perfusion, 5 = excellent. The better the perfusion reading, the more accurate the signal and number. Temperature typically doesn't affect our oximeters. But we recommend that if the reading is low, try an alternate attachment.
- If anesthetist needs to take the oximeter with the patient to another area, just press the button at the top of the monitor & the device will pop out of the module for transport use.

Anesthesia Gas Analyzer

Measures the quantity and type of gases being delivered to the patient—both inhaled and exhaled.
How to:
- Attach gas analyzing cables to monitor and patient
- Machine will then calibrate and self analyze the agent being used within 15 seconds of activation
- When changing anesthetic agents during a case, analyzer will detect the changes within a 20 second period—emphasize do not touch the screen while the analyzer detects the agent.
- Select the gas analyzer waveform on the touch screen to set the color choice & waveform display height for proper visualization.

Visualize seems like jargon.

Huh? This needs to be clearer.

Should third dash be a step?

Task 4

• • • •

Develop an Online Training Module

One of the project engineers, Terry Saussol, stops you in the hall and suggests that online training would be more effective than a user guide alone. "Anesthetists could watch a brief 'how-to' video to support the information in the user guide." After a brief conversation, you tell Terry that you will create a mock-up of the online training. E-mail Terry your plan for the online training for the instructions you wrote in Task 3. Include thumbnail sketches of the visuals, as well as a script of the audio component of the training. In your script, remember to include notes indicating when each visual should appear, when animated elements should start and stop (e.g., new text appearing, an arrow pointing to a waveform, etc.), directions to the speaker (e.g., when the speaker should emphasize specific content, pause, etc.), and any other notes that will help the technicians and speaker produce this training.

Presentation Graphics
Highlighting Important Information

The Situation

You have recently been hired as a documentation and training specialist for the Virginia Office of Group Insurance, a government office that manages state employees' medical, dental, life, and disability insurance needs. The office's mission is to provide for the negotiation, purchase, and delivery of the most competitive and cost-effective group-insurance programs for state employees and their eligible dependents. For one of your first assignments, you are asked to help the office inform state employees about recent changes to their health insurance and to provide information that will help them make informed decisions regarding their health coverage. Because Virginia is switching insurance carriers for the next fiscal year (FY), the office has been busy communicating to employees about the features and options the new insurance plans offer and how to best use the benefits.

To enroll with the new carrier, employees must complete a new enrollment form, which asks them to choose between a traditional plan and a preferred provider organization (PPO) plan and select an appropriate enrollment category (employee, employee plus spouse, etc.). They must also decide whether to enroll in a flexible spending account (FSA) for dependent-care and unreimbursed medical expenses. To best help employees make knowledgeable enrollment decisions, the Virginia Office of Group Insurance must clearly and effectively provide the information they need.

Open enrollment, the only time of the year employees may make changes to their health coverage, is approaching, and the office needs your help preparing a presentation. Open enrollment meetings will be held at several locations around the state during the month preceding the plan's open enrollment period. State employees will receive an e-mail with details regarding the meeting at their specific job location (**Figure 7.1**). These meetings will begin with an oral presentation and will conclude with a question-and-answer session.

Figure 7.1
Sample E-mail
Announcing Open
Enrollment Meeting

Subject: Invitation to Meeting on Changes to Health Benefits

In preparation for the open enrollment period next month, the Virginia Office of Group Insurance would like to invite Virginia Department of Environmental Quality (DEQ) employees to attend a presentation on upcoming changes to their insurance carrier and health plans.

Representatives from the Virginia Office of Group Insurance will hold open enrollment meetings at your facility next week on Monday from 10:00 a.m.–11:30 a.m. and on Thursday from 2:00 p.m.–3:30 p.m. Please plan to attend one of these sessions. Both meetings will be held in the DEQ second floor conference room.

The Challenge

Experienced speakers, trainers, and salespeople know that presentation graphics improve their audience's learning, retention, and perception of the presenter. These graphics also reduce the time needed to explain complex subjects. Your challenge is to support a presentation on complicated information with effective graphics so that state employees with diverse needs and concerns can make effective health coverage decisions.

Your Job

To download electronic copies of documents in this chapter, click on **Case 7 Documents** on bedfordstmartins.com /techdocs.

With your background in technical communication, the Virginia Office of Group Insurance is relying on you to help create presentation graphics that clarify and highlight important features of the new insurance plan. You may be asked to do the following:

▶ Design presentation slides.

▶ Simplify concepts and visual representations on presentation slides.

▶ Create a presentation suitable for online viewing.

▶ Prepare a handout to accompany a presentation.

Your instructor will tell you which of the tasks you are to complete.

Get started on your job (see page 76).

When You're Finished

Reflecting on This Case   In a 250- to 500-word response to your instructor, discuss (a) what you learned from this case, (b) how you could relate this case to work situations you will face in your chosen career, and, if applicable, (c) the ways in which this case compares to similar situations you have already faced at work. Your instructor will tell you whether your response should be submitted as a memo, an e-mail, or a journal entry, or in a different format.

Moving beyond This Case   Prepare a 10-minute presentation on important policies at an organization with which you are familiar, and include appropriate graphics. You might choose a place of employment, a community group, a campus club, or a sports team. Your audience consists of new members in the organization, and your purpose is to inform them of key policies that they need to know to be successful and safe within the organization.

Task 1

Design Presentation Slides

Burt Krebs, Virginia State Insurance Manager, has created a few slides (**Documents 7.2** to **7.4**) for the open enrollment meeting presentation using a popular presentation-graphics application. "I didn't want to use one of the templates because most didn't seem to offer a design that would best suit our information, but now I'm not sure how to design my slides from scratch," he explains.

DOCUMENT 7.2

Opening Slide for Presentation

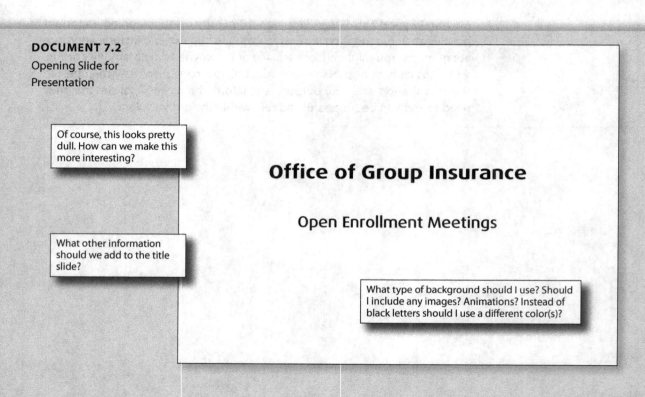

Burt hands you printouts of the slides with notes on them and says, "Take a look at these and come up with an effective design. The notes explain some of my concerns and suggest some options for design." Decide on an overall design for Burt's presentation, and apply this design to the slides Burt showed you. Then, in a brief e-mail to Burt, justify your design decisions.

DOCUMENT 7.3

Presentation Slide with Agenda

How should we improve the design to help our audience understand each slide, the organization of our presentation, and the point the speaker is addressing?

What information, if any, should we place in a footer?

Agenda

Insurance Carrier Selection
    Employee Health Insurance Survey

FY 2012 Appropriation

Transition of Care

FY 2012 Plan Design

Open Enrollment Procedures
    Online and Paper Applications

DOCUMENT 7.4

Presentation Slide on the
Paper Application

Click to add title

Review PPO Provider Directories and if Provider NOT on
PPO List, Decide if PPO is Appropriate.

Download and print a PDF of Application on Insurance
Website. Or, complete your application entirely online.

All Employees Must Re-enroll in Health Insurance
Make a Choice between Traditional or PPO

Completed Paper Application must be turned in to Agency
Payroll Office by 5:00 PM May 28.

Agency Payroll Office will enter your Information via IPOPS.

Agency Payroll Offices Will Receive a Limited Number of
Paper Applications. Supplies are limited.

This slide describes what
state employees need to
do to complete the paper
application (although we
prefer that they complete
the online application
instead). I need a title. Also,
how can we design this
slide to make these steps
clearer?

Task 2

Present Information Visually

Burt shows you printouts of three slides (**Documents 7.5** to **7.7**) used during an earlier in-house presentation to the Office of Group Insurance staff. "I threw these together for a quick staff meeting earlier this year. I'd like to use these in the Open Enrollment presentation, but I don't think the slides are clear enough for a general audience. Now that I look at them again, I'd be embarrassed if people outside this office saw them," Burt confides. He then asks you to evaluate the effectiveness of the slides. Using presentation-graphics software, revise these slides so that Virginia state employees can quickly and easily understand the presented information.

DOCUMENT 7.5
Presentation Slide with
Pie Chart

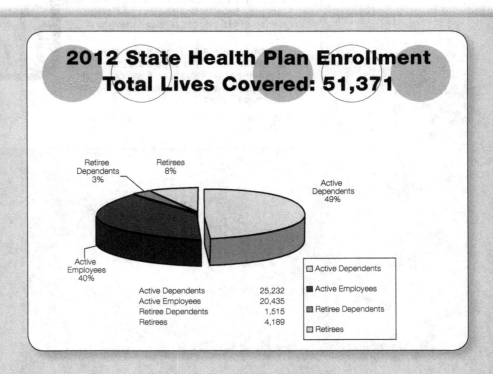

DOCUMENT 7.6
Presentation Slide on
Survey Results

Employee Survey

Concerns

- **Salary Issues.** Respondents pointed to no across-the-board pay increases for the past three years and felt that their take-home pay has been reduced as a result of premium increases each year.

- **Picking a Plan Design.** Respondents feel they have limited choices and many don't want to pay for services they don't use.

- **Pharmaceuticals.** There was a great deal of concern about the significant increase in out-of-pocket expenses if we switched from a co-payment plan to a "co-insurance"[1] plan.

1. Participants pay a percentage of the cost of medication each time a prescription is filled rather than a co-pay (e.g., $14 generic).

DOCUMENT 7.7

Presentation Slide on Benefits

Benefit Provision for Active Employees

Benefit	Traditional Plan	PPO[1] in-network	PPO[1] out-of-network
Outpatient Physical Therapy	Plan pays 80% of allowable charges after you meet your deductible. The benefit is limited to $800 per insured per benefit period.	See Outpatient Therapy Services	
Inpatient Physical Therapy	Plan pays 80% of allowable charges after you meet your deductible. The benefit is limited to $15,000 per insured per benefit period.	Plan pays 85% of allowable charges after you meet your deductible. There is a $150,000 lifetime benefit limit.	
Temporomandibular Joint Syndrome (TMJ)	Plan pays 80% of allowable charges after you meet your deductible. The benefit is limited to $2000 per lifetime.		
Transplant Services	Plan pays 80% of allowable charges after you meet your deductible. The benefit is limited to $350,000 lifetime maximum per insured.[2]	Plan pays 85% of allowable charges after you meet your deductible.	Plan pays 70% of allowable charges after you meet your deductible.

1. There is a $350,000 lifetime maximum per insured in-network and out-of-network combined.
2. There is an additional $5,000 living expenses benefit.

Task 3

• • • •

Create an Online Presentation

One important decision state employees must make during the open enrollment period is whether to choose a traditional or PPO health plan design. Because the state has never offered a PPO plan, Burt wants to answer the questions, "What is a PPO plan, and how does it work?" Consequently, he has decided not only to discuss this at the meetings but also to create a brief online presentation with audio to accompany the slides. "Although we will discuss this topic at the open enrollment meetings, I want an online resource available to state employees who either can't make the meeting or need a refresher before they select a plan next month," Burt explains. "I don't want a bunch of text-heavy slides for viewers to read. Instead, I want the presentation graphics and audio component to work together to provide a clear explanation."

Burt has e-mailed you his notes (**Document 7.8**) describing what the presentation should cover. Create four to six slides, and write the script for the audio portion of the presentation. In your script, remember to include notes indicating when each slide should appear, when animated elements should start and stop (e.g., new text appearing, an arrow pointing to important content, etc.), directions to the speaker (e.g., when the speaker should emphasize specific content, pause, etc.), and any other notes that will help the technicians and speaker produce this presentation.

DOCUMENT 7.8

Presentation Notes

To make an informed choice between a traditional plan and a PPO, state employees need to understand what a PPO is. A Preferred Provider Organization (PPO) is a health plan with an established provider network (physicians, hospitals, labs, etc.) that has agreed to accept specific payment levels for specific services.

I want to make this point clear: A PPO is not a Health Maintenance Organization (HMO). A PPO does not require you to select a Primary Care Physician (as in HMOs), and referrals are not required (as in HMOs). PPOs are an increasingly common plan offered by employers. One benefit of a PPO is that patients can "self-refer" themselves to specialists without having to get a referral from their primary care physician (PCP). You just make an appointment with the specialist. In a PPO, you are free to select any healthcare provider. If your physician is in the PPO network, you will have lower deductibles, a higher reimbursement of expenses, and the lowest possible out-of-pocket expense. If enrolled in a PPO and your physician is not in the PPO network, you will experience a higher deductible, a lower reimbursement level for expenses, and a higher annual out-of-pocket maximum limit.

As part of this conversation on PPO plans, I'd like to summarize the FY2012 plans for the audience. (The following applies to both the traditional and PPO plans.) Pharmacy benefits as well as Dental and Vision are unchanged. New for FY '12 is a premium category (Employee + Spouse + Single Child). The flexible spending account (FSA) limit has been *increased* from $2,500 to $3,000, the Integrated Behavioral Health Plan (IBHP) In-Patient Co-payment has been *increased* to $15, and Outpatient Co-payment has been *increased* to $25. Finally, enhanced Wellness Benefits and Well Baby Nursery Benefit are part of both plans.

For the traditional plan, employee premiums will increase by 15%, traditional plan benefits are unchanged (with the exception of the *addition* of Wellness and Well Baby Nursery Benefits), and co-insurance is unchanged at 80/20 of allowable charges (that is, the insurance carrier pays for 80% of the allowable charges and the employee pays for the remaining 20%).

For the PPO plan, the network is slightly smaller than the traditional network, monthly premiums and deductibles are *less* than the current FY '11 traditional insurance plan, the co-payment is $20 for physician office exams with in-network physicians, and the co-insurance is 85/15 of allowable charges (in-network) and 70/30 out-of-network.

Task 4

• • • •

Prepare a Handout

The presenters for the open enrollment meetings will be Abbey McLean and Liana Anderson. They have decided that they would like to supply attendees with a handout. "We don't want to merely print our presentation slides. We'd like a brief handout that focuses on our 2011 enrollment numbers, the difference between the new PPO and our traditional plans, benefit provisions, plan costs, and our contact information." From her smartphone, Abbey e-mails you the plan costs and their contact information (**Document 7.9**). Create a handout that can be printed on a single sheet of standard paper.

DOCUMENT 7.9

E-mail with Information for Presentation Handout

If attendees have questions, have them contact one of the following members of the Office of Group Insurance Services Team:

Liana Anderson, Benefits Assistant (L–R), 378-3170, landerson@ogi.gov

Abbey McLean, Benefits Assistant (A–K), 378-3739, amclean@ogi.gov

Trey Summerfield, Benefits Assistant (S–Z), 378-2044, tsummerfield@ogi.gov

Below are the plan costs for FY '12:

PPO Plan: Employee only $15; employee and spouse $38.50; employee and child $26; employee and children $35.50; employee, spouse, and child $48.00; employee, spouse, and children $54.50

Traditional Plan: Employee only $18.50; employee and spouse $47; employee and child $32.50; employee and children $43; employee, spouse, and child $58.50; employee, spouse, and children $65.50

Note: The above rates detail the bimonthly premium rates for employees employed at least 26 hours per week. Premiums are deducted from the employee's first and second paychecks each month. No separate premiums for vision benefits for all members (and dependents) enrolled in medical coverage.